ECONOMIC PERSPECTIVES ON ACID DEPOSITION CONTROL

ACID PRECIPITATION SERIES
John I. Teasley, Series Editor

ECONOMIC PERSPECTIVES ON ACID DEPOSITION CONTROL

Edited by Thomas D. Crocker

ACID PRECIPITATION SERIES—Volume 8
John I. Teasley, Series Editor

BUTTERWORTH PUBLISHERS
Boston·London
Sydney·Wellington·Durban·Toronto

An Ann Arbor Science Book

Ann Arbor Science is an imprint of Butterworth Publishers.

Library of Congress Cataloging in Publication Data
Main entry under title:

Economic perspectives on acid deposition control.

  (Acid precipitation series; v. 8)
  "An Ann Arbor science book."
  Initial version of each paper was presented at the
1982 annual meetings of the American Chemical Society.
  Includes index.
  1. Acid deposition—Economic aspects—Congresses.
I. Crocker, Thomas D. II. Series.
TD196.A25E26 1983   363.7'384   83–18864
ISBN 0–250–40573–3

Butterworth Publishers
80 Montvale Avenue
Stoneham, MA 02180

10 9 8 7 6 5 4 3 2 1

Printed in the United States of America

CONTENTS

THE EDITORS

JOHN I. TEASLEY, SERIES EDITOR

John I. Teasley was a career employee of the U.S. EPA and several of its predecessor organizations. His background, both academically and throughout his professional employment, was that of an analytical chemist. His duties included research functions as well as serving in line management.

He was quite extensively involved in EPA's acid precipitation program from its inception until his retirement in July 1983. He continues to remain active in the ACS's Division of Environmental Chemistry and currently serves as the Councilor for the Lake Superior Section.

THOMAS D. CROCKER, VOLUME EDITOR

Currently Professor of Economics at the University of Wyoming, Thomas D. Crocker has previously served on the faculties of the Universities of California and Wisconsin. A native of the State of Maine, he received his A.B. from Bowdoin College, and his Ph.D. in agricultural economics from the University of Missouri. He is a former member of the U.S. Environmental Protection Agency's Science Advisory Board and several National Academy of Sciences committees and panels. Author or co-author of more than eighty technical publications, his major research interests are the development of methods to value non-marketed, particularly environmental, goods, and the properties of alternative allocation systems.

THE CONTRIBUTORS

Richard M. Adams
Department of Agricultural
    Economics
Oregon State University
Corvallis, Oregon

Scott Atkinson
Department of Economics
University of West Virginia
Morgantown, West Virginia

R.A. Cabe
Department of Economics
University of Wyoming
Laramie, Wyoming

Alan Carlin
Office of Policy Analysis
U.S. Environmental Protection
    Agency
Washington, D.C.

Thomas D. Crocker
Department of Economics
University of Wyoming
Laramie, Wyoming

Ralph C. d'Arge
Department of Economics
University of Wyoming
Laramie, Wyoming

L.S. Eubanks
Division of Economics
University of Oklahoma
Norman, Oklahoma

Bruce A. Forster
Department of Economics
University of Guelph
Guelph, Ontario, Canada

Allen V. Kneese
Resources for the Future
Washington, D.C.

Fredric C. Menz
Department of Economics
Clarkson College of Technology
Potsdam, New York

John K. Mullen
Department of Economics
Clarkson College of Technology
Potsdam, New York

James L. Regens
Institute of Natural Resources
University of Georgia
Athens, Georgia
On leave to the U.S. Environmental
    Protection Agency
Washington, D.C.

John T. Tschirhart
Department of Economics
University of Wyoming
Laramie, Wyoming

SERIES PREFACE

These volumes are a result of a symposium on Acid Precipitation held in conjunction with the American Chemical Society's Las Vegas meeting, held in the Spring of 1982. The symposium was organized along nine thematic areas including meteorology, chemistry of particles, fogs and rain, oxidation of SO_2, deposition both wet and dry, terrestrial effects, aquatic effects, geochemistry of acid rain, economics, and predictive modeling.

The thematic areas were planned and conducted by expert investigators in each of the particular disciplines. The investigators were: Chandrakant Bhumralker, meteorology; Jack Durham, chemistry of particles, fogs and rain; Jack Calvert, oxidation of SO_2; Bruce Hicks, deposition; Rick Linthurst, terrestrial effects; George Hendrey, aquatic effects; Owen Bricker, geochemistry; Thomas Crocker, economics; and Jerald Schnoor, modeling.

The symposium was designed to present findings on atmospheric movement of the precursors of acid precipitation; atmospheric chemistry and precipitation effects including aquatic, terrestrial and geochemical; the economics involved with "acid rain" and finally mathematical modeling in order that future problems may be predicted.

As one studies this series it becomes readily evident that the objectives of the symposium were indeed met and that the series will serve as a ready reference in the field of acid precipitation.

I wish to acknowledge the efforts and dedication of the volume editors for the superb job they did in bringing this series to fruition. Working with people such as these made all the time and energy spent well worth while.

I further wish to thank the American Chemical Society, The Division of Environmental Chemistry and the Council Committee of Environmental Improvement for providing the facilities for the symposium from which this series evolved.

Finally, I wish to express my gratitude to the individual chapter authors without whom the series would never have been possible.

John I. Teasley

PREFACE TO VOLUME 8

In the past 20 or so years, we have been assaulted by a revival of stern Malthusian prophecies about the declining capacity of North American and world natural resource stocks to satisfy human wants. As usually advanced, the prophecies refer solely to the adequacy of stocks of industrial raw materials such as minerals, water, food and fiber, and, most recently, energy. Debates about the predictive accuracy of the prophecies revolve on whether rates of "resource-expanding" technological advance can be rapid enough to offset declines in the qualities and quantities of raw material assets available for fabrication. Except in some of the popular, nontechnical literature, the debates have infrequently confronted the fact that the resource expansions the new technologies permit have been progressively accompanied by pollutants, found not at all or in much lesser amounts in nature, having the potential to draw down the pleasures and the life support services that the earth's ecosystem scaffolding can provide. That is, to soften the perceived problem of materials depletion, we could be depleting the ability of natural ecosystems to supply life support services. Acid deposition, which primarily originates in various energy-producing technologies, is said to be capable of contributing to this ecosystem depletion. Parts of our ecosystem scaffolding are being washed in a continuing series of acid baths. Many scientific and political authorities believe that irritated places have appeared.

As the President's Council on Environmental Quality stated in its 1979 annual report, environmental issues are inevitably economic issues. If one group is to be made better off, another will often be worse off. Acid deposition, one of the most contentious environmental issues to strike the public consciousness in the last decade, is no exception. The topics of the chapters in this volume bound the domain of acid deposition issues on which economic methods can offer insight. Treatments of the economic value of natural science information, the benefits and costs of controlling precursors, the involuntary exchange problems that interregional and international transport pose, decision-making under uncertainty, the incentive properties of alternative control instruments, and the ethical presuppositions necessary for policy applications of the results of economic studies are all present in one or more chapters.

Competitively organized markets convey information about relative scarcities sufficient to provide incentives to individuals to allocate goods to whom, where, and when they will be most highly valued. All exchanges that can make some group better off without harming someone else are identified and voluntarily implemented. The physical and institutional facts of the acid deposition phenomenon frustrate this competitive market outcome. Its achievement requires some use of the police power of the state. Each of the chapters in this volume deals with some facet of generating the information that public policymakers could use to bring about the outcomes that would otherwise result from competitively organized markets in rights to cause and to inhibit acid deposition.

An initial version of each paper was presented at the 1982 annual meetings of the American Chemical Society. The cross-disciplinary good will of the Society members, particularly Dr. John I. Teasley, is enlightened. After presentation, each paper was revised one or more times to take account of written comments by fellow volume authors and a group of outside readers that included Richard Bishop of the University of Wisconsin, Ronald Cummings of the University of New Mexico, Jack Donnen of the Ontario Ministry of the Environment, J. Wayland Eheart of the University of Illinois, and Edward Jesse of the University of Wisconsin.

Mrs. Carol Steadman has ably typed and edited the format of several of the papers, and has also assisted in arranging logistical details of the original presentations and the construction of this volume. I alone have responsibility for any lapses that may remain in the volume.

Thomas D. Crocker

CHAPTER 1

Introduction

Alan Carlin

To my knowledge this session represents a first—the first time that a technical seminar has been held on the economics of controlling acid deposition. Since I assume that fossil fuel use will continue at least at something approaching current levels, it seems likely that this will only be the first of many such seminars. This seminar is also somewhat unusual in that it brings together what is largely a group of economists at meetings sponsored by a leading professional organization in the physical sciences. I can only applaud the breadth of view represented by this decision. It is particularly appropriate in this case because of the broad, interdisciplinary nature of the acid deposition problem. It is quite clear that economists cannot make much progress on it without the help of physical scientists. Similarly, the ultimate political decisions that may be made as to whether and how to address the problem will require both physical science and economic inputs.

This introduction will touch on two areas: (1) why the Office of Research and Development (ORD) of the U.S. Environmental Protection Agency (EPA) believes the subject to be of some importance; and (2) a brief introduction to the papers to be presented.

IMPORTANCE OF THE ECONOMICS OF ACID DEPOSITION

The economics of acid deposition are of potential importance from two legal/procedural viewpoints. The first is that, pursuant to Executive Order 12291, proposed regulations judged to be "major" must have Regulatory Impact Analyses (RIA) prepared on them that include a cost/benefit analysis. The second is that under current law (although there has been some discussion in Congress of possible changes in the Clean Air Act to provide for specific regulation of acid deposition), one of the more feasible approaches to regulating acid deposition is through use of a Secondary Standard. One possible interpretation of this section of the statute is that the requirement that the standard protect the "public welfare" may in part imply some sort of economic efficiency test.

Executive Order 12291, signed February 17, 1981, less than a month after the Reagan Administration came into office, does not alter the legal basis on which

1

EPA must make its decisions under the environmental laws. It does, however, require that these RIA be prepared for the information of the decision-makers and subject to review by the Office of Management and Budget. It is too early to say just how important these RIA will prove to be, but they do for the first time make a cost/benefit analysis a part of the formal decision-making process (although not a legal basis for most decisions).

In general, economic efficiency considerations do not play a prominent role as criteria for decisions under the Clean Air Act as currently written. In fact, one of the few sections where such an interpretation is possible is that for Secondary Standards. Because there are few indications that acid deposition has major adverse health effects, the Secondary Standard approach may be one of the more applicable sections under the present act. If so, the issue will obviously be what constitutes protecting "public welfare." If it includes promoting economic efficiency, then this topic will be of some importance from a regulatory standpoint. Let me hasten to add, however, that I am not a lawyer and cannot speak for the EPA General Counsel's Office, so these semilegal views of mine should not be taken as Gospel.

However, even if the law should be changed or interpreted differently or if the Executive Order should be cancelled, there will always remain the intellectually important question of whether possible control actions are expected to have net positive or negative effects on national economic efficiency, or if one adopts a broader perspective suitable to this international issue, on the economic efficiency of the affected region without respect to national boundaries. Whether or not economic efficiency criteria are a formal part of any regulatory process that may be undertaken in the future, reasonable people will want to know whether proposed actions are likely to have net positive or negative efficiency effects.

For all these reasons, ORD has funded a limited research effort on the economic benefits of controlling acid deposition as part of its much larger, continuing effort to develop improved methodology and data for determining the economic benefits of pollution control. The results of this research form the basis of several of the papers presented here. Our purpose in funding this early acid deposition research has not been to find the exact dollar cost of acid deposition damages, although that might indeed ultimately be useful, but rather to assess current knowledge, solve some of the methodological issues that must be resolved before much progress can be made toward that end, and suggest what physical and biological research will provide the greatest dividends for improving our understanding of the economic benefits of control. Clearly, if we want to determine these benefits, we will make more progress if we can focus that portion of physical and life science research that may be devoted to the end of determining economic benefits on those areas with the largest economic effects.

SURVEY OF PAPERS

Now that I have made the argument that the economics of acid deposition control are of some interest, let me proceed to the second part of this introduction, a brief survey of the papers to be presented. In looking through the papers, I find it most useful to group them into five general topics: policy perspectives, economic benefits of control, international aspects, economically optimal control strategies, and economic caveats.

The policy perspectives topic is covered in Chapter 2 by James Regens. His chapter centers on the current level of scientific knowledge in relation to possible control decisions.

The second area, economic benefits of control, or alternatively, the reduced damages of acid deposition, is represented by three chapters. Richard Adams and Thomas Crocker (Chapter 4) discuss economically relevant dose-response estimation and the value of information. Frederic Menz and John Mullen (Chapter 9) assess the economic damages caused by acid deposition in the most publicized area of acid deposition impact, the Adirondack fisheries. Bruce Forster (Chapter 7) discusses the economic damages caused by acid deposition from a Canadian perspective.

Before proceeding, it may be important to explain the difference between benefits, costs and damages for those who do not deal with these concepts every day. By benefits, I and most economists mean the positive economic effects of control actions; by damages I mean the adverse economic effects of pollution, in this case acid deposition. Obviously, the benefits of a control action are equal to the damages prevented by the action. On the other hand, economic costs, as used here, refer to the costs of implementing pollution controls. Benefits may be more or less than costs; the difference is called the net economic benefit of a control measure. Executive Order 12291 sets out the policy that subject to legal constraints net benefits should be positive.

The third area is represented by a chapter on the legal, economic and political aspects of transfrontier pollution by two well known environmental economists, Allen Kneese and Ralph d'Arge (Chapter 8). The acid deposition problem is perhaps the best known example of such pollution problems because of its long range transport character.

The fourth area, comprising three papers, is on the general topic of optimal control strategies, which has been a major area of study with respect to other pollution problems. Because of the unusual local vs distant trade-offs found in the case of acid deposition, it is of particular interest here. The first two papers in this category, by Scott Atkinson (Chapter 3) and John Tschirhart (Chapter 10), deal directly with this issue. Although they do not directly relate to the benefit area, they would obviously be of importance if an economic approach to control were ever to be adopted. The third paper, by Thomas Crocker (Chapter 5) considers the relation between scientific findings and policy choices under uncertainty, a topic of particular relevance for acid deposition.

The last category, which I have called caveats, contains one paper by Larry Eubanks and Richard Cabe (Chapter 6) that outlines the many assumptions and other caveats that should be taken into account in attempting to apply and interpret the results of applying cost/benefit analysis to problems such as this.

SOME FINAL THOUGHTS

Although the chapters do not provide all the answers we are going to need to make economically sensible decisions about controlling acid deposition, they do raise a number of the issues that must be addressed. I believe that we need to pay particular attention to the economic benefits and costs of control not only to meet the procedural requirements but also to ensure that whatever decisions are ultimately made about control are made with the best possible knowledge of their economic consequences. Although the costs of control need to be better understood, the benefits represent the largest gap in our current knowledge. To fill this gap will require the combined efforts of the physical, biological and economic sciences. The chapters presented here represent a significant start in this direction by both summarizing current knowledge and suggesting where further physical and biological data are needed. Many of the gaps in our knowledge of the benefits of control are a result of inadequate physical or biological data; improved knowledge of the economic data needed to determine the benefits of control can point the way towards collecting physical and biological data that will later allow the economic estimates to be improved.

CHAPTER 2

Acid Rain: Does Science Dictate Policy or Policy Dictate Science?

James L. Regens

As Victor Hugo once said, "Science says the first word on everything, and the last word on nothing." The topic of "acid rain" clearly illustrates this point. Its causes and effects have been the subject of substantial claims and counter-claims in the last several years. Such controversy points to the conclusion that this specific environmental problem is subject to major scientific uncertainty. For example, without structural understanding of the processes underlying the transport and chemical transformation of precursor emissions into acidic compounds and the ultimate deposition of these compounds, it is not possible, with any significant degree of precision, to match a given overall reduction in emissions of specific pollutants to an overall reduction in resultant acidity levels. In addition, the lack of a well defined source-receptor relationship (the fundamental transport question) precludes certainty as to whether reductions from one geographic locale will produce desired outcomes in another, particularly reductions in episodic or seasonal as opposed to annual average deposition rates. As a result, our current ability to estimate the success or failure of intervention strategies such as SO_2 emission reductions producing desired results is severely limited.

Why does this uncertainty exist? In a broad sense, for each research problem, there is some range of uncertainty about the choice of appropriate research strategies, the measurement and data analysis techniques themselves, and the findings produced by the research. To the extent that there are scientific uncertainties associated with individual hypotheses, those uncertainties expand substantially when looking at complex research questions. Such a situation characterizes the current state of knowledge with respect to the nature, extent and causal factors underlying the effects attributable to acid deposition.

OVERVIEW OF THE PROBLEM: A SCIENTIFIC PERSPECTIVE

The "acid rain" problem stems from of a series of complex and varied chemical, meteorological and physical interactions. The term "acid rain" (itself somewhat imprecise) commonly refers to the transfer of acidic materials, whether wet (precipitation) or dry (deposition), from the atmosphere onto the earth's surface with resultant impacts on natural or manmade structures and objects. By convention, precipitation is considered "acidic" if the collected solution has an acidity greater than water saturated with atmospheric carbon dioxide (i.e., pH $<$ 5.6). The definition is somewhat arbitrary, since evidence of acid rain in remote areas has been observed at several monitoring sites. Natural as well as anthropogenic sources may influence some of these sites. Thus, the concept of a global background of pH 5.6 appears questionable.

Unfortunately, the simplicity of the term "acid rain" also conveys the image of an easily measured and understood phenomenon. In fact, the material that falls to earth from the atmosphere is quite complex, and its components are measured accurately only with extreme difficulty [1]. Although the hydrogen ion (H^+), or "acidity" is the most obvious and visible feature of "acid precipitation," it should be emphasized that this particular attribute results from the presence of a host of contributing chemical species which are derived from both natural and manmade sources. Those positively and negatively charged ionic constituents most important in this regard are the anions Cl^-, SO_4^{2-}, SO_3^{2-}, NO_3^-, PO_4^{2-}, OH^- and CO_3^-, and cations NA^+, K^+, NH_4^+, C^{2+}, Mg^{2+} and H^+.

It is important that analyses of precipitation chemistry trends not focus solely on H^+. So that all the source species can be considered for meaningful relationships to be established, measurements of precipitation chemistry should include all species contributing to the total ion balance. This often has not been the case, and since all species to be measured are in very low concentration, there is considerable potential for error. Often, sufficient data do not exist to compute a complete ion balance and determine acidity. This can be remedied to some extent by using approximations derived from monitoring network data, but these tend to be highly geographic-specific and must be applied with caution. For example, in the Midwest and Great Plains, dry seasons and land-use practices may increase the concentrations of soil components in precipitation and decrease the resulting acidity. As a result, by necessity, the study of acid deposition includes questions about the role of local vs distant sources of emissions, the formation and transport of acidic compounds within the atmosphere, their deposition as dry or wet material, and the effects induced by these materials as they move through an environment which has differing capabilities to absorb acid deposition.

Current Levels and Trends

Because good methods are not available for measuring total (wet + dry) acid deposition, monitoring programs have focused on the precipitation (wet) component. Dry deposition measurements generally have been limited to intensive studies conducted with highly sophisticated equipment and specialized conditions, but the error bars associated with these assessments are quite large, impeding understanding of impacts on receptors. In particular, dry deposition rates of SO_x, NO_2, sulfate, nitrate and HCl over various terrains and seasons of the year must be determined to understand the relative contributions and importance of wet and dry deposition to the acidification of lakes, streams and other components of the environment.

Continuous, large-scale precipitation monitoring has been conducted in Scandinavia since the 1950s, and similar monitoring was conducted on a limited basis in the United States before the mid-1970s. However, with the exception of the Hubbard Brook Experimental Forest in New Hampshire, consistent North American data are available for less than a decade. As a result, long-term trends in the North American context are defined poorly. Attempts by Likens et al. [2] to piece together fragmentary precipitation data to demonstrate increasing acidity in eastern North America over the past 20 years have been criticized by others using the same data due to the limitations in the methods used to derive estimates of acidity levels in the 1950s [3,4].

Recent measurements, however, indicate that large areas of eastern North America as well as portions of the western United States are subject to quite acidic precipitation on an annual average [5,6]. "Summer" and "winter" seasonal distributions for pH are similar, but the East has substantially more acid deposited in the summer, while seasonal patterns are different for much of the West. This variability suggests the importance of seasonal rainfall in determining total acid deposited in precipitation. The chemistry of the constituents also varies seasonally, increasing by a factor of two or three for sulfate content of precipitation in much of the East during the summer [3]. Variability between storms and even within storms also is documented. Many individual storm events in the Northeast have rainfalls of pH < 4.0. For example, Johannes et al. [7] report that for a six-month period in 1978, 31 rainfall events had pH < 4.0 while 17 events were greater than 4.0 in the Adirondack lake region of New York.

Transport, Transformation and Deposition

The principal acids found in precipitation samples collected in eastern North America are sulfuric acid and nitric acid [3]. Substantially lesser quantities of hydrochloric acid also are usually observed. Predicting transport, transformation and deposition of these precursor emissions appears particularly complex [7,8]. Several mechanisms contribute to deposition and the chemical quality of precipitation, and many of these steps tend to be reversible in nature. Pollutant species tend to be highly interactive, and multiple species often must be considered simultaneously in evaluating deposition mechanisms and impacts. Emissions of both sulfur and nitrogen oxides can be transformed into compounds that are more acidic than the original state. Such transformations, which may involve photochemical oxidants and other air pollutants, can occur during atmospheric transport, in cloud and rain droplets, or after deposition to ground or water. Factors such as water vapor content, stack height, temperature, sunlight, windfields, precipitation and ground cover can influence the distance that emissions are transported before deposition. This makes it extremely difficult to be able to predict, on the basis of emission sources, spatial and temporal distributions for the atmospheric deposition of NO_3^-, SO_4^{2-}, and other important chemical species.

Although modeling should not be expected to substitute for monitoring in defining source-receptor relationships, modeling can be used in a broad sense to express mathematically individual or composite processes in the deposition sequence. Understanding of atmospheric deposition processes is reflected, in large measure, by the ability to model validly and reliably those phenomena. In most cases, site-specific estimates of precipitation chemistry obtained on the basis of deposition climatology for that receptor area have greater reliability than do those derived from model calculations. Finally, because field measurements in the form of network monitoring are essential for model development, a strong interactive component exists in the modeling-monitoring relationship.

Until recently, however, the question of long-range transport had received little emphasis in the context of conventional air pollution modeling. Two major problems have combined to limit the accuracy and reliability of trajectory analysis techniques:

1. inadequacy of wind data to produce trajectory calculations having acceptable accuracy; and
2. uncertainties regarding vertical motions and pollutant distributions tend to detract from the accuracy of trajectory calculations.

Consequently, there is a strong need for better resolution of routine wind observations, more comprehensive archiving of wind measurements, and continued progress in modeling of tropospheric flow fields before really adequate input for trajectory calculations can be achieved. In principle, many of the problems associated with trajectory calculation can be eliminated by transport calculation via more complex modeling approaches. These are difficult to perform in practice because of the necessity for highly involved numerical analyses and large amounts of computational time. An additional concern of trajectory calculations is their verification through experimental measurement. Trajectory verification efforts will have a strong positive effect on future capabilities to predict deposition on the basis of specific area sources. Specifically, tracer studies are especially important for the development and verification of long-range transport models in order to delineate source-receptor relationships.

Much remains to be accomplished in this important research area. Present modeling capabilities are constrained by four major deficiencies: (1) limited species representation; (2) inadequate characterization of transformation chemistry; (3) inadequate characterization of interphase transport of particulate material; and (4) inadequate representation of complex flow fields. No existing model includes all of these considerations in a physically realistic manner. Until the knowledge base supporting that capability is improved, modeling has limited utility for assessing the benefits of emission restrictions by quantitatively predicting reductions in acid deposition.

The uncertainties of long-range transport model outputs also apply to the uncertainties in the elements of transfer matrices. A transfer matrix is an array of numbers linearly relating sources to receptors. Using a transfer matrix, deposition at a given site from a set of emittors can be estimated by multiplying the appropriate element in the matrix by the emission. This procedure is only as good as the reliability of the matrix elements, and the validity of the assumption that deposition is proportional to emissions. Consequently, before transfer matrices are used to estimate source-receptor relationships, more reliable data and changes in input parameters are required.

Environmental Effects

Concern about acid rain results largely from studies that indicate that aquatic and terrestrial ecosystems, and materials may be affected adversely by acidification [9,10]. However, the extent and magnitude of the problem, its likelihood of occurrence in a given area, and the implications of the effects are not well established. Comprehension of the aquatic processes potentially altered by acidic deposition and its effect on aquatic resources is based primarily on statistical associations between acidic deposition patterns, trends in surface and groundwater chemistry, and associated changes in biological status. Chemical changes in surface water quality and associated declines in some species of fish and invertebrates are the best documented effects [11–16]. Interference with normal reproductive processes in acidified waters can also be induced by increased concentrations of certain cations, notably aluminum, through toxic effects on egg and larvae stages [17–19].

Acidification of lakes and streams, and a reduction in fish populations have been observed in the northeastern United States, southern Ontario and Quebec, Scandinavia, and southwestern Scotland [20–23]. Such apparently simple correlations are the result of many interacting processes that are currently understood only superficially. Acidification of lakes and streams may occur through both long-term depletion of alkalinity and episodic events that involve a sudden increase in hydrogen ion concentration. Such episodic events are caused by the rapid release of contaminated snow melt or from acidic rainstorms [2,14,24,25]. Common to these areas are granite or other bedrock with little or no buffering capacity; thin and patchy soils; poorly buffered surface waters; and the presence of acidic precipitation [2]. As a result, the relative contribution of natural and anthropogenic inputs to acidification is unclear [26].

Impacts on terrestrial ecosystems (e.g., agricultural crops, forests) are of major concern due to the high economic and recreational value of these resources. Although direct effects of acidic deposition on natural vegetation have not been observed under ambient field conditions, laboratory studies using simulated acid rain suggest that changes in nutrient availability and nutrient cycling may affect the productivity of vegetation [27–29]. Such changes may modify plant nutrient budgets or alter the resistance of plants to drought and disease stresses as well as influence in both a stimulatory and inhibitory manner the growth of forest seedlings [30,31]. The indirect effects of acid precipitation potentially include effects on soils. The impacts of acid precipitation on soils vary greatly, however, depending on the characteristics of the ecosystems in which they are found [32–34].

Recently, debate also has developed relative to the release of aluminum from soils to aquatic ecosystems [18,35,36]. Aluminum released as a result of soil acidification similarly has been postulated as a possible cause for death and a decline in growth of feeder roots in spruce, fir and beech [37]. While it is well documented that aluminum is toxic to vegetation, the relationship between acidic deposition and aluminum toxicity is less certain.

Because of their complexity, specification of the impacts of acidic deposition on natural ecosystems will not be obtained easily. Substantial questions remain to be resolved, including delineating which criteria are to be employed in order to assess adequately the degree to which terrestrial resources are at risk from acidic deposition. For example, an understanding of the effects of acid deposition on the leaching of nutrients from vegetation and soils, the consequences of excess sulfur, nitrates, and potentially toxic metals, microbial processes, and the productivity of unmanaged terrestrial ecosystems are some of the outstanding scientific issues which need to be quantified.

The impacts of acid precipitation on crops is another major area of scientific uncertainty [38–41]. Given the economic value of agriculture, a small change in productivity has the potential to result in a significant net loss. Current knowledge based on laboratory experiments, however, is inadequate to support a conclusion that ambient acid precipitation does or does not significantly affect the yield (quantity and/or quality) of commercially grown crops in either a statistical or economic sense. Responses are subtle and include stimulatory, inhibitory and null effects. For some crops, all three types have been noted, apparently depending on the experimental design and environmental conditions.

Data supporting adverse impacts by precipitation on manmade structures and materials are also very limited. The most applicable findings from experimental studies of degradation due to dry deposition, SO_2, ozone and particulate matter suggest that surface corrosion by SO_2 and its acid derivative is dependent on various environmental factors, including, relative humidity or the duration of surface wetness [42,43]. In order to develop meaningful economic assessments of the damages resulting from these degradation processes and potential mitigation measures, the role played by acid deposition must be distinguished from that of nonacidic, gaseous pollutants and natural weathering of materials.

Finally, available information provides no conclusive evidence of direct health effects from acid precipitation. However, indirect risks could arise if acidified waters result in increased levels of toxic elements, such as mercury in fish or heavy metals in drinking water supplies. Although potentially correctable through treatment actions, some cases of drinking water drawn at the tap exhibit levels of metals (lead, copper, zinc) at concentrations higher than U.S. drinking water standards [44,45]. Further research, especially inventories of water supply systems, is needed to determine whether and the extent to which impacts on drinking water supplies are linked to acid deposition.

RESEARCH AND POLICY: LINKING SCIENCE TO DECISION-MAKING

Addressing the acid rain question requires coordination of a complex research effort that at this stage, primarily has identified the relatively inchoate status of our existing knowledge base. The reduction of current scientific and technical uncertainties requires analytical contributions from a variety of disciplines, each of which have their own measurement techniques and approaches. If scientific information is to provide guidance about policy options, the findings from these disciplines then must be integrated in a policy-relevant fashion. In addition, the acid deposition problem involves "real world" uncertainties about relationships over a variety of spatial and temporal horizons. Relatively good understanding of how changes in one variable affect others in a laboratory does not necessarily equate with a capability to predict how these variables interact in the real world. Finally, developing a policy response to the acid rain problem involves assessments of relative risks in situations where we not only lack good historical data, but we are not even certain that the baseline for our data will be sufficient for assessing future risks and consequences.

While the above discussion focuses on the state of existent knowledge, two types of uncertainty exist for the policymaker seeking to link scientific information to policy choice. The first is essentially of a scientific nature. In the realm of science, one deals with a body of knowledge and range of uncertainty about the results that can be expected due to alterations in physical, biological or behavioral contexts. To the extent that the researcher can control sources of variation, it is possible to establish relatively unambiguous relationships with clear confidence levels. For a number of environmental problems, such as acid deposition, where the mechanisms are extremely complex, it is clear that significant commitments of research talent, time and financial resources are necessary before the causal linkages underlying effects can be delineated adequately enough to identify cost-effective responses to prevent or mitigate effects. In such situations, the policymaker faces a second and critical kind of uncertainty. This latter uncertainty consists of an estimate of the value of further information prior to action, based on the expected gains from additional research, relative to the damages that might be incurred in the interim as well as the cost of proposed actions if they turn out to be ineffective.

The degree to which policymakers are willing to accept uncertainty and still take action rather than adopt a "wait and see" approach depends on several things. First, it depends on estimates of the magnitude and reversibility as well as the value of damages and the likelihood that those damages will occur. Second, it depends on the cost of taking action and the probability that the option implemented fails to remedy the problem. One of the few areas characterized by relatively little uncertainty is the question of the costs of some of the proposals commonly put forward to reduce acid deposition. The economic cost of reducing SO_2 emissions is approximately \$300–480 per ton removed [46]. Because such expenditures represent a substantial commitment of society's resources, the opportunity cost of being wrong is a reduction in the money available for controls in the future. However, in the event that some of the effects on ecosystems are irreversible, inaction also represents an opportunity cost in the form of damages which could not be mitigated. In a sense, this provides a conceptual framework within which policymakers can weigh regulatory responses to the problem of acid deposition.

At this time, there is clearly a gap between what is known and what ideally ought to be known from a scientific perspective before society imposes further restrictions on ambient emissions. While science can aid in policy choice, it would be naive to assume that decision-making is exclusively a function of scientific knowledge. Moreover, regulatory action is not always explicitly or implicitly based on the assumption that the public sector has established, with reasonable certainty, the existence of a specifiable degree of actual or potential damage. Similarly, knowledge, whether complete or parital (the latter being the current case with respect to acid deposition), of a phenomenon's occurrence does not necessarily provide understanding of the factors that cause it to occur. However, without such understanding, the ability to predict the efficacy of interventions designed to alter such an occurrence is dubious at best. Even given the willingness to spend "X" dollars to reduce a specific risk, because damage presumably would equal or exceed that amount, the merits of allocating those dollars without some identifiable degree of confidence that the risk will be diminished (i.e., actual benefits would accrue) is questionable from a normative standpoint.

In essence, deferring the imposition of additional control measures theoretically provides the opportunity for learning about the control benefits that the status quo offers while continued research in the physical and biological sciences can better define the parameters of the problem. Adopting such a viewpoint (i.e., the position of the Reagan administration) rests on the assumption that additional research (i.e., science) can help provide insight as to whether regulatory action actually would affect the problem, presumably in a cost-effective way. Although our ability to make policy decisions is hampered by unresolved scientific questions, to the extent that controls are found to be needed in the future or if significant damages are irreversible, then the cost savings from delaying abatement will be illusory. However, even if the latter scenario proves to be accurate (i.e., irreversibility of damages), it remains unclear what specific control strategies would be efficacious, since the quantitative physical and chemical relationships between SO_2 and NO_x emissions and a given acidic deposition level are unavailable [47]. Thus, information has the potential to inform judgment not only as to the nature of the problem but also to identify the options available for addressing the problem.

EFFORTS TO REDUCE UNCERTAINTY

In the face of existing levels of scientific uncertainty, Congress enacted the Acid Precipitation Act of 1980, passed as Title VII of the Energy Security Act. This legislation instructed the Environmental Protection Agency (EPA), in cooperation with the other members of the Interagency Task Force on Acid Precipitation, to identify specifically the causes, sources and effects of acid precipitation through an extensive research effort. To this end, Congress established a ten year, comprehensive interagency research plan. The Task Force serves not only to coordinate research at the federal level, but also cooperate with other levels of government [48]. Based on such analyses, regulatory action should be taken to the extent necessary and practicable to prevent or otherwise ameliorate harmful effects attributable to acid deposition.

This gives rise to the question: how is the EPA specifically (and the overall Task Force in general) committing its resources to reduce the gap between what is known and what ideally might be known before the federal government can successfully mitigate this problem? For fiscal year (FY) 1982, EPA has committed $9 million out of the approximately $18 million being devoted by the federal government to the study of the acid deposition problem (Table I). A significant part of EPA's research effort is the National Trends Network. This network will provide consistent data on acid deposition, which will not only permit the examination of trends in acid deposition, but it also will be possible to define more precisely the geographic extent of potential effects because of increasingly acidic deposition. Deposition monitoring will receive approximately $800,000 during FY 1982, and $1.1 million in FY 1983 (Table II), to assure the quality and accuracy of data collected under this system. These data collection efforts will be augmented by a global trends network to look at transboundary trends, and a research support network which will provide special study sites.

On the effects side, EPA is spending approximately $3 million in this fiscal year ($3.8 million proposed for FY 1983) to examine the effects of acid deposition on aquatic and terrestrial species and manmade materials and to assess naturally occurring effects. These studies have as their ultimate goal assessing levels of damage and establishing dose-response relationships.

Another crucial element of the research program is the attempt to better understand atmospheric processes. This involves a variety of questions, including concerns about the meteorological mechanisms underlying the mechanical movement of materials by wind, chemical and physical transformation processes, atmospheric suspension times, and the chemical nature of deposited material. Projects in this area will account for over $3.2 million in FY 1982 and $3.8 million in the FY 1983 research budget. These efforts will help to determine what mechanisms might influence the formation, transport, and deposition of acidic substances. Naturally, it is important to evaluate potential sources of precursor emissions. Although both natural and manmade sources play a role, assessments of their relative contributions provide insights into not only the magnitude of those contributions, but of how changes in them may affect acid deposition.

Table I. FY 1982 Budget ($1000s) for the Interagency Task Force on Acid Precipitation

Task Group	EPA	USDA SE	USDA FS	NOAA	DOE	DOI PS	DOI USGS	DOI FW	TVA	Total
Natural Sources				600						600
Manmade Sources	870				300					1,170
Atmospheric Processes	3,273			600	805	50			135	4,863
Deposition Monitoring	797	304	127	700	497	75	585		18	3,103
Aquatic Impacts	1,475		135		560	98	645		154	3,067
Terrestrial Impacts	1,355	964	570		122	186	320		66	3,583
Effects on Materials	250					235				485
Assessments and Policy	1,105				260					1,365
Subtotals		1,268	832			644	1,550	0		
Totals	9,125	2,100		1,900	2,544		2,194		373	18,236[a]

[a] Resources for control technology (Task Group H) are not included in this budget.

Table II. President's FY 1983 Budget ($1000s) for the Interagency Task Force on Acid Precipitation

Task Group	EPA	USDA		NOAA	DOE	DOI			Total
		SE	FS			PS	USGS	FW	
Natural Sources				700					700
Manmade Sources	1,050				300				1,350
Atmospheric Processes	3,858			1,050	600	50			5,558
Deposition Monitoring	1,145	404	145	500	340	100	1,449		4,083
Aquatic Impacts	1,935		210		550	148	445	75	3,363
Terrestrial Impacts	1,468	946	1,258		50	229	486		4,437
Effects on Materials	410					535	50		995
Assessments and Policy	1,570				220				1,790
Subtotals		1,350	1,613			1,062	2,430	75	
Totals	11,436	2,963		2,250	2,060		3,567		22,276[a]

[a] Resources for control technology (Task Group H) research and development are not included in this request.

Evaluations of mitigation strategies also form an important part of this research effort. Whether it is more cost-effective to control at the source of emissions, or to mitigate at the effects end, depends on the relative dose-response functions and costs of different control options. These questions can be addressed once more information is available about the sources and how changes in those sources will affect deposition.

CONCLUSION

Additional knowledge is necessary to delineate properly the causes and effects of acid deposition. For that reason, research and monitoring efforts are important. While a reasonable amount of knowledge about precursor emissions does exist, the situation with regard to the transport and transformation of these precursors is much less certain. It is clear, however, that there are significant differences in the composition and acidity of rain between seasons, in different geographical areas and even between storms. At this time, whether summer sulfate levels are influenced mostly by the amount and types of oxidizing agents in the atmosphere or the amount of SO_2 in the atmosphere is unknown. Similarly, ambiguities exist over why remote regions have rains with low pH. Because knowledge of the distances pollutants may travel is changing rapidly, understanding of the conversion processes which enable acid deposition to form is limited. Hence, assumptions about the role of specific pollutants as well as comparisons of emission sources must be done within the context of clear knowledge constraints about how pollutants are transported and transformed. Finally, it appears that the potential welfare effects attributable to deposited acid materials may include impacts on manmade materials; high elevation lakes and headwater streams; crops; forests and other natural ecosystems; and certain soil systems. However, the extent as well as the mechanisms by which acid deposition is linked causally to actual effects has not been specified with any real degree of precision. As a result, for both SO_2 and NO_x, adequate models for identifying source-receptor relationships with sufficient precision to permit a comprehensive assessment of control options are unavailable at present.

Research can contribute significantly to reducing the current level of uncertainty about the causes and consequences of the acid deposition problem. However, science can never provide the last word with absolute certainty. Clearly, such a mission represents an impossible task, although research can be used to improve significantly the quality of information in areas important for decision-makers. Enhanced understanding of the causal factors and potential consequences of acid deposition enables policymakers to identify and select those courses of action which, given their own value systems, address most appropriately the problem. Thus, science neither dictates the policy nor policy the science. Instead, a symbiotic relationship exists between the two. Knowledge from one domain can foster choice in the other.

REFERENCES

1. Tyree, S.Y. "Rainwater Acidity Measurement Problems," *Atmos. Environ.* 5:57–60 (1981).
2. Likens, G.E., F.H. Bormann and N.M. Johnson. "Acid Rain," *Environment* 14:33–40 (1972).
3. Semonin, R.G., V.C. Bowerson, D.F. Gatz, M.E. Pedan and G.J. Stensland. "Study of Atmospheric Pollution Scavenging (May)," #DE-ACOZ-76EV01199, U.S. Department of Energy, Washington, DC (1981).
4. Hansen, D.A., and G.M. Hidy. "Examination of the Basis for Trend Interpretation of Historical Rain Chemistry in the Eastern United States," ERT Document NO. P-A097, Environmental Research and Technology, Inc. (1981).
5. Liljestrand, H.M., and J.J. Morgan. "Chemical Composition of Acid Precipitation in Pasadena, California," *Environ. Sci. Technol.* 12:1271–1273 (1978).
6. Stensland, G.J. "Precipitation Chemistry Trends in the Northeastern United States," in *Polluted Rain,* T.Y. Toribara, et al., Eds. (New York: Plenum Publishing Corporation, 1980), pp. 87–104.
7. Johannes, A.H., E.R. Altwicker and N.L. Clesceri. "Characterization of Acidic Precipitation in the Adirondack Region," EA-1826, Electric Power Research Institute, Palo Alto, CA (1981).
8. Junge, G.E. *Air Chemistry and Radioactivity* (New York: Academic Press, Inc., 1963).
9. Oden, S. "The Acidification of Air Precipitation and Its Consequences in the Natural Environment," Swedish National Science Research Council, Ecology Commmittee Bulletin #1, Stockholm (1968).
10. Overrein, L. "Final Report and Bibliography of the SNSF Project," SNSF Project, Oslo, Norway (1981).
11. Schofield, C.L. "Effects of Acid Precipitation on Fish," *Ambio* 5:228–230 (1976).
12. Beamish, R.J., and H.H. Harvey. "Acidification of the LaCloche Mountain Lakes, Ontario and Resulting Fish Mortalities," *J. Fish. Res. Board Can.* 29:1131–1143 (1972).
13. Dickson, W., C. Ekstrom, E. Hornstrom and U. Miller. "Forsurningens inverkan pa vastkustsjoar" (The Effects of Acidification of West Coast Lakes), Publ. No. 7, Swedish National Environmental Protection Board, Stockholm (1973).
14. Leivestad, H., G. Hendrey, I.P. Muniz and E. Snekvik. "Effects of Acid Precipitation of Freshwater Organisms," in "Impact of Acid Precipitation on Forest and Freshwater Ecosystems in Norway," F.H. Braekke, Ed. Research Report FR 6/76, (Oslo, Norway: SNSF-Project, 1976), pp. 86–111.
15. Henriksen, A. "Acidification and Freshwaters—A Large Scale Titration," in *Ecological Impact of Acid Precipitation,* D. Drablos and A. Tollan, Eds. (Oslo, Norway, SNSF Project, 1980), pp. 68–74.
16. Fritz, E.S. "Potential Impacts of Low pH on Fish and Fish Populations," FWS/OBS-80/40.2 U.S. Fish and Wildlife Service, Biol. Serv. Prog. Nat. (1980).
17. Hultberg, H., and O. Grahn. "Proceedings of the First Specialty Symposium on Atmospheric Contributions to the Chemistry of Lake Waters," *J. Great Lakes Research 2 (Supplement 1)* (1976).
18. Driscoll, C.T., J.P. Baker, J.J. Bisogni, Jr. and C.C. Schofield. "Effects of Aluminum Speciation on Fish in Dilute Acidified Waters," *Nature* 284:161–164 (1980).
19. Grahn, O., "Fish Kills in Two Moderately Acid Lakes Due to High Aluminum Concentration," in *Ecological Impact of Acid Precipitation,* D. Drablos and A. Tollan, Eds. (Oslo, Norway; SNSF Project, 1980), pp. 310–311.

20. Leivestad, H., and I.P. Muniz. "Fish Kill at Low pH in A Norwegian River," *Nature* 259:391–392 (1976).

21. Henriksen, A. "A Simple Approach for Identifying and Measuring Acidification of Freshwater," *Nature* 278:542–545 (1979).

22. Likens, G.E., R.F. Wright, J.N. Galloway and T.J. Butler. "Acid Rain," *Sci. Am.* 241(4):43–51 (1979).

23. Harvey, H.H. "The Acid Deposition Problem and Emerging Research Needs in the Toxicology of Fish," in *Proceedings of 5th Annual Aquatic Toxicology Workshop,* Hamilton, Ontario, Fish. Mar. Serv. Tech. Rep. #862, (1979), pp. 115–128.

24. Johannessen, M, A. Skartveit and R.F. Wright. "Streamwater Chemistry Before, During, and After Snowmelt," in *Ecological Impact of Acid Precipitation,* D. Drablos and A. Tollan, Eds., (Oslo, Norway: SNSF Project, 1980), pp. 224–225.

25. Galloway, J.N., C.L. Schofield, G.R. Hendrey, N.E. Peters and A.H. Johannes. "Sources of Acidity in Three Lakes Acidified During Snowmelt," in *Ecological Impact of Acid Precipitation,* D. Drablos and A. Tollan, Eds. (Oslo, Norway: SNSF Project, 1980), pp. 264–265.

26. Last, F.T., G.E. Likens, B. Ulrich and L. Walloe. "Acid Precipitation—Progress and Problems," in *Ecological Impact of Acid Precipitation,* D. Drablos and A. Tollan, eds., Oslo, Norway, SNSF Project, 1980), pp. 10–12.

27. Wood, T., and F. H. Bormann. "Effects of an Artificial Acid Mist on the Growth of Betula Alleghaniensis," *Brit. Environ. Poll.* 7:259–268 (1974).

28. Shriner, D.S. "Effects of Simulated Rain Acidified with Sulfur Acid on Host-Parasite Interactions," in *Proc. First Int. Symp. on Acid Precipitation and the Forest Ecosystem,* L.S. Dochinger, Ed., USDA, For. Serv. Gen. Tech. Rep. NE-23, Upper Darby, PA, (1978), pp. 919–925.

29. Shriner, D.S., and J.W. Johnston. "Effects of Simulated Acidified Rain on Nodulation Leguminous Plants by *Phizobiven* spp.," *Environ. Exp. Bot.* 21:199–209 (1980).

30. Lee, J. J., and D.E. Weber, "The Effect of Simulated Acid Rain on Seedling Emergence and Growth of Eleven Woody Species," *Forest Sci.* 25:393–398 (1979).

31. Raynal, D.J., A.L. Leaf, P.N. Maion and C.J. Wang. "Actual and Potential Effects of Acid Precipitation on Forest Ecosystems in the Adirondack Mountains," Report 80–28, New York Statement Envir. Research and Development Authority (1980).

32. Hutchinson, T.C., and L.M. Whitby. "Heavy Metal Pollution in the Sudbury Mining and Smelting Region of Canada. 1. Soil and Vegetation Contamination by Nickel, Copper and Other Metals," *Environ. Conserv.* 1(2):123 (1974).

33. McFee, W.W. "Effects of Acid Precipitation and Atmospheric Deposition on Soils," in *A National Program for Assessing the Problem of Atmospheric Deposition (Acid Rain),* J.N. Galloway, et al., Eds., Nat. Atm. Dep. Prog., Nat. Res. Ecol. Lab., Fort Collins, CO (1978), pp. 64–73.

34. McFee, W.W. "Sensitivity of Soil Regions to Acid Precipitation," U.S. EPA Report, EPA-600/3–80–013 (1980).

35. Cronan, C.S., and C.L. Schofield. "Aluminum Leaching Response to Acid Precipitation: Effects on High-Elevation Watersheds in the Northeast," *Science* 204:304–305 (1980).

36. Erikson, E. "Aluminum in Groundwater; Possible Solution Equilibria," *Nordic Hydrol.* 12:43–50 (1981).

37. Ulrich, B., R. Mayer and P.K. Khanna. "Chemical Changes Due to Acid Precipitation in a Loess-Derived Soil in Central Europe," *Soil Sci.,* 130:193–199 (1980).

38. Evans, L.S., C.A. Conway and K.F. Lewin. "Yield Responses of Field-Grown Soybeans

to Simulated Acid Rain," in *Ecological Impact of Acid Precipitation*, D. Drablos and A. Tollan, Eds., (Oslo, Norway: SNSF Project, 1980).

39. Jacobson, J.S. "The Influence of Rainfall Composition on the Yield and Quality of Agricultural Crops," in *Ecological Impact of Acid Precipitation*, D. Drablos and A. Tollan, Eds., (Oslo, Norway: SNSF Project, 1980).

40. Evans, L.S., and K.F. Lewin. "Growth, Development, and Yield Responses of Pinto Beans and Soybeans to Hydrogen Ion Concentrations of Simulated Acidic Rain," *Environ. Exp. Bot.* 21:103–113 (1981).

41. Lee, J.J., and G.E. Neely. "CERL-OSU Acid Rain Crop Study Progress Report," Corvallis Environmental Research Laboratory (1981).

42. Kucera, V. "Effects of Sulfur Dioxide and Acid Precipitation on Metals and Anti-Rust Painted Steel," *Ambio* 5:243–248 (1976).

43. Haynie, F.H. "Theoretical Air Pollution and Climate Effects on Materials Confirmed by Zinc Corrosion Data," paper presented at the First International Conference on Durability of Building Materials and Components, Ottawa, Ontario, 1979.

44. Fuhs, G.W., and R.A. Olsen. "Acid Precipitation Effects on Drinking Water in the Adirondack Mountains of New York State," Technical Memorandum, New York State Department of Health Laboratories, Environmental Health Center, Albany, NY (1979).

45. Fuhs, G.W. Statement Before the Subcommittee on Environment, Energy and Natural Resources, Committee on Government Operations, U.S. House of Representatives, (16 July, 1980).

46. ICF, Inc. "Analysis of Alternative NSPS Regulations," Draft Report, Washington, DC (1981).

47. Evans, L.S., G.R. Hendrey, G.J. Stensland, D.W. Johnson and A.J. Francis. "Acidic Precipitation: Considerations for an Air Quality Standard," *Water, Air Soil Poll.* 16:469–509 (1981).

48. "First Annual Report to the President and the Congress of the United States," Interagency Task Force on Acid Precipitation, Washington, DC (1982).

CHAPTER 3

Effect of Global Optimization on Locally Optimal Pollution Control: Acid Rain

Scott Atkinson

Despite efforts for more than a decade to improve ambient air quality, most urban areas appear unable to attain the 1982 federal ambient standards for ozone, sulfur oxide, particulate matter, nitrogen oxide, and carbon monoxide. In fact, many major urban areas are projected to be in nonattainment for more than one of these pollutants [1]. In addition, the extent of acid deposition, not currently governed by standards, has increased in the Midwest and Northeast over the last decade [1]. If noncompliance with standards is to be avoided, the U.S. Environmental Protection Agency (EPA) will have to increase individual source control requirements or Congress will have to relax either the ambient standards or their deadlines.

Economists have proposed alternative "least-cost" strategies for achieving ambient air quality standards at minimum total regional cost. They require specifying source pollution control requirements, imposing pollution taxes or allowing the trade of marketable pollution rights. The magnitude of these requirements, taxes or marketable pollution rights must be adjusted so that the ambient standards are just achieved at all air quality receptors. Individual control requirements would minimize control costs if the marginal control costs per unit of ambient improvement were equated across all sources at the level which satisfied ambient standards. Taxes on ambient degradation could also be designed to achieve identical results, since each firm will equate the tax to its marginal cost of control. Finally, the trading of marketable pollution rights for ambient degradation, recently considered by EPA [1], would achieve equivalent results in competitive markets without administrative and political constraints on the trading of permits. Each firm would equate the price of marketable permits to its marginal cost of control. Crocker [2] gives an early recognition of this point and Montgomery [3] presents a detailed mathematical proof of this proposition.

This chapter demonstrates, however, that least-cost strategies for SO_2 control that meet only local standards will most likely lead to increased long-range SO_4 deposition, more popularly termed "acid rain." This increase is due to the greater

21

degree of local environmental loading with SO_2 emissions and to the incentive for sources with taller stacks to undertake less control under the local least-cost strategies than the current system of air quality control. Thus, without the imposition of constraints on interregional SO_4 transport, the adoption of local least-cost air quality strategies will only worsen the present extent of SO_4 deposition. Therefore, this chapter attempts to measure the magnitude of this phenomenon for the Cleveland region and to examine the implications of SO_4 deposition constraints on the cost advantage of the local least-cost systems.

Local least-cost strategies are basically variants of either emissions least-cost (ELC) or ambient least-cost (ALC) control strategies. Under the ELC control strategy, sources with the lowest marginal control costs are required to undertake the greatest control burden, independent of their degradation of local air quality. Under the ALC control strategy, sources with the lowest marginal control costs per unit improvement of local air quality are required to undertake the greatest control. That is, cost-effectiveness of incremental local control in achieving ambient standards determines control responsibility. As discussed above, both strategies could be implemented via transferable pollution permits, standards, or charges for either emissions or ambient degradation.

Both the ALC and ELC local least-cost strategies have been modeled using mathematical programming techniques, in St. Louis by Atkinson and Lewis [4], where particulate control was considered, and in Chicago by Anderson et al. [5], where nitrogen dioxide control was examined. Both studies indicate an order-of-magnitude reduction in local control costs under the ALC system and a fivefold reduction under the ELC system compared to the current system based on State Implementation Plans (SIP).

However, both least-cost strategies result in greater loading of the environment with SO_2 emissions than the SIP strategy. First, the latter is not a cost-minimizing strategy and therefore requires more control to achieve the ambient standard than the local least-cost strategies. Second, the ALC strategy generally encourages sources with tall stacks (usually power plants) to undertake less control, since they degrade the local environment less than other sources. Thus, the potential for long-range residual transport is much greater under the local least-cost strategies, whose economic incentives for adoption appear very strong.

The present study attempts to measure the extent to which the costs savings of the local ALC strategy is due to externalizing the costs of long range SO_4 deposition. Sulfur oxide control is examined for the Cleveland region of the Ohio River Basin. Since the Cleveland area is projected to be a nonattainment region for sulfur oxides, schemes for assigning additional control requirements are examined. A quadratic mathematical program is used to model the ALC method of pollution control. The objective function is a quadratic cost function, reflecting increasing marginal control costs. When only local ambient standards are satisfied, a local or intraregional ALC solution is obtained. By adding constraints on the long-range transport of SO_4, an interregional ALC solution is also obtained. Clearly, the interregional solution is more costly than the intraregional ALC solution, since some externalities are internalized. Inter- and intraregional strategies that allocate additional control in proportion to current SIP control requirements are also examined. The first assigns additional control until local ambient standards are met. The second assigns additional control to meet these standards plus constraints on long-range SO_4 transport.

The results indicate that the intraregional ALC strategy is substantially more cost-effective than the intraregional SIP strategy. However, this saving is due to cost-effective pollution control, which implies greater loading of the local environment with SO_2 and substantially reduced control of power plants with tall stacks. Both contribute heavily to long-range transport of SO_4. Once significant constraints on long-range transport are introduced into both strategies, the cost savings of the interregional ALC strategy is substantially reduced relative to the interregional SIP strategy.

MODEL FORMULATION

This chapter compares the costs and control burdens of four strategies. The intraregional strategy, which includes only local SO_2 ambient constraints, is modeled for both the ALC and SIP methods. The interregional strategy, which includes local SO_2 ambient constraints and long-range SO_4 transport constraints, is also modeled for both the ALC and SIP methods.

The intraregional SIP allocation strategy requires computing additional control requirements for each source in proportion to its percentage contribution to total regional uncontrolled emissions under the current SIP strategy. The additional emission control responsibility for each source is computed as

$$AER_j^e = AER^e \frac{ue_j}{\sum\limits_{j} ue_j}, \qquad j = 1, \ldots, n \qquad (1)$$

where AER^e = required change in aggregate regional emissions
         ue_j = uncontrolled emissions from the jth source under the current strategy

The value of AER^e necessary to achieve the 24-hour maximum SO_2 standard is adjusted iteratively, after an initial best guess, by computing individual source AER_j using Equation 1 and running uncontrolled source emissions through a diffusion model to determine expected air quality at each receptor. Adjustments in AER^e are then made in successive iterations until the required ambient standards are just met at the region's air quality receptor yielding the highest pollutant reading. Since there is no guarantee that this receptor will remain the same in each iteration, the diffusion model results must be examined at each iteration.

The interregional SIP strategy requires, in addition, that the following constraint on long-range SO_4 transport be satisfied:

$$\sum_l a_l \sum_j x_j \geq b \tag{2}$$

where a_l = long-range transport coefficient mapping SO_2 emissions from Cuyahoga County into $\mu g/m^{-3}$ of SO_4 in the lth neighboring region
x_j = SO_2 to be removed (g/s^{-1}) by the jth source
b = total required reduction in SO_4 concentrations ($\mu g/m^{-3}$) in all regions
due to Cuyahoga County SO_2 emissions

We model the intraregional ALC control method by assuming that the cost
functions for the sources under study are nonlinear, convex and continuously twice
differentiable and that all firms minimize their cost of control given the requirement
that they install scrubbers. We can write the cost-minimization problem for the
intraregional ALC method as a quadratic programming (QP) problem:

$$\text{minimize } z = \sum_j c_j x_j + \sum_j d_j x_j^2 \tag{3}$$

$$\text{subject to} \quad \sum_j a_{ij} x_j \geq b_i, \qquad i = 1, \ldots, n, \tag{4}$$

$$x_j \geq 0, \qquad j = 1, \ldots, n, \tag{5}$$

where c_j, d_j = coefficients representing the cost of control per day for the jth
source ($j = 1, \ldots, n$)
a_{ij} = transfer coefficients that relate emissions from the jth source to
air quality at the ith receptor
b_i = reduction in SO_2 concentration required to achieve the maximum
24-h SO_2 standard at the ith receptor

If u_i = uncontrolled air quality at receptor i and g is the air quality standard,
$u_i - g = b_i$, i = 1, . . . , m. If marketable pollution permits are employed, this
trading rule requires the purchase of an ambient permit by each source for each
receptor plus monitoring by the control agency and possible adjustment of the
number of permits to guarantee that ambient standards are met.

Finally, the interregional ALC control method can be written as the QP
problem in Equations 3–5 plus the additional acid deposition constraints in Equation 2.

DATA AND ALGORITHMS

This section describes the Cleveland area source inventory, the local SO_2 air quality
diffusion model and the long-range SO_4 transport model, and the control cost
algorithms, which are input to the quadratic programming algorithm employed
in this study. The quadratic programming algorithm is SYMQUAD [6], based
on the work of Van de Panne and Whinston [7].

Source Inventory

The sources examined in this study comprise the 25 largest point-source emittors of SO_2 in the Cleveland area defined by Cuyahoga County. Data on these sources and their emission characteristics was obtained from Robert Hodanbosi, Chief of the Division of Air Quality Modeling and Planning of the Ohio EPA. Together these point sources account for 95% of total regional emissions from all major point sources and 60% of total ambient degradation. The remaining ambient degradation from excluded point sources and all area sources is treated as background. The facility name, SO_2 source, and percent SO_2 content of the fuel burned by each source are given in Table I. Of the 25 sources, 14 generate electricity, 1 source is a steam-heat boiler, 1 a wastewater treatment plant, and the remainder are industrial sources. As seen in Table II, utility generating sources account for 94.85% of SO_2 emissions and these sources employ substantially taller stacks than the other sources. Thus, the potential for long-range acid deposition is great for these emitters.

Local Diffusion Model and Long-Range Transport Model

Air quality diffusion modeling is performed using the RAM steady-state Gaussian dispersion model developed by the EPA [8]. Principally, RAM determines from 1-hour to 1-day urban air quality concentrations resulting from pollutants released by point and area sources. The algorithm is applicable for locations with level or gently rolling terrain, where a single wind vector, mixing height and stability class are assumed representative of the entire area for each hour.

The following information is required for all point and area sources. Emission information for all point sources includes source coordinates, emission rate, physical stack height, stack diameter, stack-gas exit velocity and stack-gas temperature. Emission information required of area sources consists of location coordinates, source side-length, total area emission rate and effective area source height. The output of the RAM model consists of calculated ambient concentrations at each receptor for hourly averaging times. Computations are performed on an hour-by-hour basis as if the atmosphere had achieved a steady-state condition. Therefore, if concentrations gradually build up over time, as under light wind conditions, error may result. The total concentration for a given hour at a particular receptor is the sum of the estimated contributions from each source.

Source-receptor pollutant transfer coefficients are computed from the RAM-calculated contributions of each source to the ambient concentrations at each receptor by dividing each concentration in micrograms per cubic meter by the emission rate of the contributing source in grams per second. The RAM model was run to simulate a typical 24-h interval. The maximum 24-h SO_2 standard is 365 $\mu g/m^{-3}$. Thus, the output of the diffusion model was examined to determine whether this standard was violated during any of the 24 h. The hour and receptor associated

Table I. Inventory of SO$_2$ Sources in Cuyahoga County

Source	Source Number	SO$_2$ Process	Fuel	Percent SO$_2$
ALCOA	1,2	Industrial coal boiler	Coal	3.57
Cleveland Electric Illumination Co.	3	Steam heat boilers	Coal	0.70
Avon Lake Plant	4–7	Electricity generation	Oil	0.24
	8–10	Electricity generation	Coal	2.65
Eastlake Plant	11,12	Electricity generation	Coal	3.14
Lakeshore Plant	13–16	Electricity generation	Oil	1.62
	17	Electricity generation	Coal	0.74
Division Pumping Station	18	Water Treatment plant	Coal	2.24
Ford Engine Plant No. 2	19	Coal boilers	Coal	2.9
General Motors, Chevrolet	20	Coal boilers	Coal	2.26
Lincoln Electric	21	Coal boilers	Coal	2.9
Medical Center	22	Coal boilers	Coal	1.88
Republic Steel	23	Boiler	Coke oven gas	
	24	Slab RF No. 1	Coke oven gas	
	25	Slab RF No. 2	Coke oven gas	

Table II.   Potential Long-Range Transport Under Existing Controls

Source[a]	Current Emission Rate (g/s^{-1})	Stack Height (m)
1	106.80	60.96
2	72.20	45.72
3	79.60	95.71
4–7	148.90	84.40
8	1,578.90	152.40
9	1,668.00	119.50
10	4,501.30	182.90
11	5,067.40	163.00
12	4,986.00	182.80
13–14	453.81	81.69
15–16	521.88	81.69
17	1,055.16	81.69
18	40.80	69.49
19	329.51	41.76
20	35.44	27.43
21	44.60	18.59
22	174.60	58.52
23	145.50	68.60
24–25	68.40	49.10
Total	22,569.6	

[a] See Table I for complete descriptions of sources.
[b] Data given are for each source, not total.

with each violation as well as the extent of each violation are given in Table III. The required improvements in air quality at each of the eight receptors in violation are employed as the right-hand side of the ambient air quality constraints in the QP solutions to the ALC problem.

The long range transport model, described in "Matrix Methods to Analyze Long-Range Transport of Air Pollutants [9], was developed by Brookhaven and Pacific Northwest National Laboratories. Separate matrices were computed for utility sources (assumed stack height of 200 m) and industrial sources (assumed stack height of 100 m) for July meteorology, which corresponds to the month of the RAM diffusion model run. The matrix coefficients map millions of metric tons of SO_2 emitted per day in one state into average concentrations in micrograms per cubic meter of SO_4^{2-} in a receptor state. In constraint Equation 2, employed by the interregional least-cost strategy, long-range transport coefficients obtained from "Matrix Methods to Analyze Long-Range Transport of Air Pollutants" [9] are converted into units of $\mu g/m^{-3}/g^{-1}/s^{-1}$. The total contribution of the 25 Cuyahoga County sources to other regions' SO_4^{2-} concentrations is computed as 13.33 $\mu g/m^{-3}$ for all utility sources and 0.79 $\mu g/m^{-3}$ for all industrial sources. In the

Table III. Receptors Exceeding the Maximum 24-h Standard

Receptor	Hour	Air Quality ($\mu g/m^{-3}$)	Required Improvement to Achieve Standard ($\mu g/m^{-3}$)
1	20	600.12	235.12
2	16	366.85	1.85
6	23	366.44	1.44
9	16	443.90	78.9
14	23	439.54	74.54
19	2	546.51	181.51
22	24	662.45	297.45
26	23	375.25	10.25

following analysis of interregional strategies, the cost of reducing these contributions by 50% is computed. Biologists have specified that reductions of this amount are necessary to restore acidified lakes to normal conditions [10]. It should be recognized, however, that the exact link between emission and interregional air quality is still unsolved.

Cost Algorithms

The costs of SO_2 control have been computed by Mitre Corporation with separate algorithms for industrial and utility sources [11]. All industrial sources are assumed to be retrofitted with dual alkali flue gas desulfurization (FGD) equipment. Based on discussions with Ohio EPA officials, dual alkali systems appear more likely to be adopted than spray dryer systems. For each industrial source, the total annual cost of FGD equipment is the sum of annualized capital costs plus annual operation and maintenance (O&M) costs. Capital costs are estimated in a regression equation as a function of the actual flue gas flowrate, thermal input, coal SO_2 content and actual SO_2 removal. The annualization factor for capital costs is 0.1715. Annual O&M costs include the annual costs of electricity, water, lime and Na_2CO_3. These costs are estimated in separate regression equations as a function of SO_2 content of fuel input, percent SO_2 removal, plant capacity and resource input costs. Finally, labor, maintenance and overhead costs are computed as a function of capital costs.

All utility sources are assumed to be retrofitted with limestone FGD systems. Again, discussions with Ohio EPA officials indicated that limestone systems appear more likely to be adopted than spray dryer systems. The estimated total annual cost of FGD equipment is estimated by Mitre Corporation based on data supplied by the Tennessee Valley Authority. Assumptions on the life of utility FGD systems and capitalization factors are identical to those for industrial FGD systems. Total annualized cost has been estimated by Mitre as a quadratic function of the percentage of SO_2 removal, megawatt production and the SO_2 percentage of fuel burned.

These estimated regression equations for industrial and utility sources are then used to compute a regression equation for each source explaining total annual costs as a quadratic function of only the grams per second of SO_2 removed by each source, assuming all other variables remain constant. First, for each source the fixed values of all right-hand side variables except grams per second of SO_2 removed were substituted into the appropriate estimated regression equations defining total annualized cost. Next, the range of SO_2 removal was varied from 1 to 90%, and the corresponding values for estimated total annualized cost (ETAC) were generated. Finally, 25 separate cost functions were estimated by regressing

Table IV. Estimated Coefficient Values for Equation 6 for Significant Point Sources

Source j	$\hat{\alpha}_j$	$\hat{\beta}_j$	$\hat{\gamma}_j$	Upper Limit of x_j
1	2,509,860	6,339.53		96.12
2	1,927,430	8,117.61		64.08
3	2,360,400	7,599.95		71.64
4	830,521	−234.73	42.5649	134.01
5	830,521	−234.73	42.5649	134.01
6	847,631	−240.91	43.6850	134.01
7	847,631	−240.91	43.6850	134.01
8	2,645,850	−111.85	1.9127	1,421.01
9	2,620,730	−127.93	2.0708	1,501.20
10	11,834,300	−132.82	0.7967	4,051.17
11	20,218,200	−189.06	1.0073	4,560.66
12	20,095,200	−191.85	1.0389	4,487.40
13	1,184,110	−104.12	6.1948	408.42
14	1,184,110	−104.12	6.1948	408.42
15	1,342,410	−109.52	5.6664	469.69
16	1,342,410	−109.52	5.6664	469.69
17	1,061,900	−184.18	4.7129	949.64
18	1,416,020	11,407.00		36.72
19	5,922,820	3,203.01		296.55
20	1,459,360	12,454.70		31.89
21	1,562,750	10,811.90		40.14
22	3,822,710	4,721.41		157.14
23	3,232,670	5,247.22		130.95
24	2,641,110	8,316.42		61.56
25	2,641,110	8,316.42		61.56

the ETAC data for source j as a quadratic function of the corresponding g/s of SO_2 removed:

$$ETAC_{jk} = \alpha_j + \beta_j x_{jk} + \gamma_j x_{jk}^2 + e_{jk}, \quad k = 1, \ldots, 90, \tag{6}$$

where $x_{jk} = SO_2$ removed (g/s^{-1}) given the kth percentage of SO_2 removal by the jth source

The $R^2/$ values obtained for all sources exceeded 0.99. The estimated intercept for each source approximates the fixed cost component and the nonintercept estimated portion approximates the variable costs of additional SO_2 control. Thus, the QP objective function may be written as the sum over all j sources of the estimated variable cost portion of Equation 6.

$$z = \sum_j (\hat{\beta}_j x_j + \hat{\gamma}_j x_j^2) \tag{7}$$

Only the quadratic terms for the utility sources were significant in Equation 6. Since these terms were highly insignificant for nonutility sources, Equation 6 was reestimated with γ_j constrained to zero for these sources. The final estimates of α_j, β_j and γ_j and the upper limit of SO_2 removal in grams per second are given for each source in Table IV.

RESULTS

The individual source control requirements and total fixed and variable annual costs for each of the four strategies of pollution control, described above, are presented in Table V. The expectation was that the intraregional ALC strategy would be substantially cheaper than the intraregional SIP strategy. Further, it was expected that inclusion of constraints on long-range SO_4 deposition in both systems would substantially reduce the cost advantage of the intraregional ALC strategy. These expectations are in fact borne out by the simulation results.

First, at the intraregional level, fixed costs are constant for all strategies at $96,381,764. However, the variable costs of the ALC strategy are less than one-third those of the SIP method to satisfy the intraregional 24-h maximum SO_2 standard. This is due to the ALC strategy's allocation of control responsibility according to the lowest marginal cost per unit of local air quality improvement. The SIP method simply increases all sources' control responsibility by an equal percentage, regardless of individual source marginal cost per unit of air quality improvement.

However, this intraregional cost savings is achieved through greater long-range transport of SO_4 under the ALC strategy. This is due to two factors. First, since the ALC strategy is cost-effective on an intraregional basis, it leads to far greater intraregional environmental loading with SO_2. The ALC strategy removes

Table V. Control Responsibilities and Costs Under Alternative Control Strategies

| | Control Requirements (g/s^{-1}) | | | |
| | Emissions Allocation Method | | ALC Method | |
Source	Intraregional	Interregional	Intraregional	Interregional
1	48.06	53.40		
2	32.49	36.10		
3	35.82	36.10		
4,5	67.01	74.45	46.50	50.73
6,7	67.01	74.45	43.92	49.50
8	710.51	789.45	861.14	1,096.86
9	750.60	834.00	841.68	1,016.97
10	2,025.58	2,250.65	1,863.17	2,646.48
11	2,280.33	2,533.70	93.84	2,120.94
12	2,243.70	2,493.00	92.33	2,057.86
13,14	204.21	226.91	184.52	338.04
15,16	234.85	260.94	200.72	370.04
17	474.82	527.58	19.54	452.83
18	18.36	20.40		
19	148.28	164.76		296.56
20	15.95	17.72		
21	20.07	22.30	21.97	
22	78.57	87.30		
23	65.48	72.75		
24,25	30.78	34.20		
Total Removed	10,156	11,281	4,653	11,305

| | Total Annual Costs ($\$/y^{-1}$) | | | |
| | Emissions Allocation Method | | ALC Method | |
	Intraregional	Interregional	Intraregional	Interregional
Variable Costs	20,464,284	25,036,800	6,523,292	22,588,548
Fixed Costs	96,381,764	96,381,764	96,381,764	96,381,764

less than one-half the total SO_2 emissions controlled by the SIP method to achieve the same intraregional SO_2 ambient standards. Ultimately, this implies greater long-range transport of SO_4 by the ALC strategy. Further, from Table II it can be seen that sources 10–12 are power plants with the tallest stacks (163–183 m). Thus, these sources undertake far less control under the ALC strategy because they degrade local air quality relatively little. This also implies greater long-range transport of SO_4.

Once constraints on long-range transport are included, the intraregional cost advantage of the ALC strategy is largely eliminated. Interregional solutions for both the SIP and ALC strategies require a 50% reduction of the contribution of the 25 Cuyahoga County sources to SO_4 concentrations in all regions. After imposing this constraint, intraregional SO_2 emissions for both strategies are highly similar, and the SO_2 removal levels are almost 2.5 times larger than under the intraregional ALC solution. Further, the power plants with the tallest stacks (sources 10–12) are required to undertake substantially higher levels of control than under the ALC intraregional strategy. The costs of both strategies have risen substantially relative to their intraregional counterparts, since one-half of previous externalities are now internalized. However, the cost advantage of the ALC strategy has almost been entirely neutralized. The total variable cost of the ALC strategy is $22,588,548, while that of the SIP method is $25,036,800. It then remains to be determined whether the cost savings of the ALC system will be sufficient to offset the possible additional costs of administering the ALC system. Further, concerns about equity in changing from the SIP to the ALC strategy may prevent adoption of the ALC system even if it creates a net reduction in costs.

CONCLUSIONS

Substantial debate has recently been heard over the appropriate strategy to achieve compliance with ambient standards. The use of ALC systems has been suggested by EPA as a far more cost-effective method of achieving this goal than the current SIP strategy. This chapter demonstrates, using data on a set of 25 SO_2 sources in Cuyahoga County, that the intraregional cost savings of the ALC strategy are substantial. However, this savings is largely due to the long-range transport of locally generated SO_2. The introduction of constraints on long-range SO_4 transport substantially reduces the cost-saving advantage of the ALC strategy relative to the SIP strategy. The question then becomes whether the remaining differential is substantial enough to cover any additional administrative costs of the ALC system and outweigh concerns about equity losses.

REFERENCES

1. "To Breathe Clean Air," Report to Congress, National Commission on Air Quality, Washington, DC (1981).
2. Crocker, T.D. "Structuring of Atmospheric Pollution Control Systems," in *The Economics of Air Pollution,* H. Wolozin, Ed. (New York: W.W. Norton and Co., 1966), pp. 61–86.
3. Montgomery, D.W. "Markets in Licenses and Efficient Pollution Control Programs," *J. Econ. Theory* 5:395–418 (1972).
4. Atkinson, S.E., and D.H. Lewis. "Determination and Implementation of Optimal Air Quality Standards," *J. Environ. Econ. Manage.* 3:363–380 (1976).
5. Anderson, R.J., Jr., et al., "An Analysis of Alternative Policies for Attaining and

Maintaining a Short Term NO_2 Standard," report to the Council on Environmental Quality prepared by Mathtech, Inc., Princeton, NJ (1979).

6.   Cohen, C., and J. Stein. "Multi Purpose Optimization System Users Guide, Version 4," Vogelback Computing Center, Northwestern University, Evanston, IL (1978).

7.   Van de Panne, C., and A. Whinston. "The Symmetric Formulation of the Simplex Method for Quadratic Programming," *Econometrica* 37:507–527 (1969).

8.   Turner, B.D., and J. Hrenko Novak. "User's Guide for RAM, Volume 1. Algorithm Description and Use," U.S. EPA, Research Triangle Park, NC (1978).

9.   "Matrix Methods to Analyze Long-Range Transport of Air Pollutants," DOE/EV-0127, U.S. Department of Energy (1981).

10.   Committee on the Atmosphere and Biosphere, National Academy of Sciences. *Atmosphere-Biosphere Interactions: Toward a Better Understanding of the Ecological Consequences of Fossil Fuel Combustion* (Washington, DC: National Academy Press, 1981.)

11.   "Acid Rain Mitigation (ARM) Study: Final Utility and Industrial Boiler FGD Costs and Derived Cost Algorithms," Mitre Corporation, McLean, VA (1981).

CHAPTER 4

Economically Relevant Response Estimation and the Value of Information: Acid Deposition

Richard M. Adams
Thomas D. Crocker

This chapter deals with issues concerning the response of ecosystems to acid deposition and the applicability of such response measures to economic assessments. Specifically, the chapter (1) reviews the generation of response functions and specifies the properties of excellent dose-response information from an economic perspective; (2) develops a decision theoretic framework for assessing the *ex ante* value of response information concerning an environmental pollutant such as acid deposition; and (3) generates *ex post* estimates of the policy worth of such information on ecosystem response to a surrogate pollutant. The last objective seeks to examine the regulatory utility of alternative levels of response information as well as to assess the value of greater precision in that information. These discussions cover a wide range of bioeconomic issues, some of which have been previously addressed in efforts dealing with analytically similar economic problems. The empirical basis of the analysis derives from recent experiments conducted by the U.S. Environmental Protection Agency (EPA) and others on pollutant response.

Agricultural economists have traditionally used biologically derived parameters to perform economic analyses of production-oriented decisions [1]. This history has led to a fairly complete statement of economically relevant experimental design and data needs that should be forthcoming from the biologist [2]. The conditions under which the response parameters should be addressed, the economically relevant range of parameter estimates and the value of additional observations in increasing the precision of biological parameters are reviewed below.

Increasing the precision of a given parameter estimate requires the acqustion of more observations on the underlying physical relationship. However, acquiring information is not costless. The more complex the structural set of explanatory variables, the more difficult parameter estimation becomes, *ceteris paribus,* with the complexity of response relationships increasing as one moves from managed monocultures associated with agriculture to naturally occurring ecosystem equilibria. While more observations will reduce the variance of an estimate, the value of those observations must be weighed against the cost of acquisition.

To assess the value of (additional) information within the context of a response relationship, the appropriate measure of benefit (or loss) and associated function must be specified. For a profit-maximizing entity, the benefits function is typically defined in terms of profits or returns. In this setting, the gain or loss in profits to the individual from additional information on a given response measure can be viewed as the value of the additional information. For public policy actions, however, the form or specification of the objective function is more uncertain. From a normative perspective, one criterion may be to compare the cost of increased information on ecosystem response with associated increases in economic surplus resulting from improved policy recommendations. Under this criterion, the value of (1) obtaining *any* information on ecosystem response and (2) obtaining increased precision in such a response parameter from additional observations can be calculated in terms of changes in economic surplus for alternative policy actions.

Employing these concepts, we present a formal framework for measuring the *ex ante* value of information about acid deposition-response relationships and then empirically derive *ex post* estimates for a case of ambient oxidant impacts upon an ecosystem. In view of the current status of acid deposition effects research, the *ex ante* structure appears appropriate. Oxidant (ozone) response information on agricultural crops then provides an empirical example of the decision process applied to a surrogate pollutant. The oxidant analysis, while admittedly exploratory, also suggests some criteria for the efficient allocation of biological research resources where economic assessments are of eventual concern.

DOSE-RESPONSE FUNCTION

The concept of a dose-response function is central to the assessment of biological or physical damages associated with pollution. It is somewhat but not perfectly analogous to the economist's production function, which has traditionally been the source of input-output information used by agricultural economists to assess farm level decision problems. Issues such as the appropriate measure of dose, the algebraic form of the function to be estimated and the measure of response must typically be addressed. These relationships are also important to the economist, who must integrate dose-response information into an economic framework to assess economic consequences. If the economist is to use dose-response information generated by the biologist in an economic assessment, then the economist must suggest criteria to guide the acquisition of this information.

Production Function

The economist's concept of the production function often differs in subtle but economically important ways from the natural scientist's idea of a dose-response function. The conceptual basis is well developed in most intermediate production economics texts; for the more advanced reader, Henderson and Quandt [3] may be of interest. Key concepts in providing guidance to the natural scientist concern the delineation of the economically relevant region of the response or production surface and the possibility for and extent of input substitution in the production process. Each concept derives from the behavior of the region of positive marginal productivities of the inputs. Thus, to the extent that acid deposition affects the positive marginal productivities of inputs such as fertilizer or water, exogenous environmental factors are potentially important shifters of biological and, hence, economic optima. Inclusion of an environmental stress factor within the structure of the production function can indicate the existence of substitution possibilities as mitigative strategies, which are important in calculating economic benefits of controlling pollution. These and other specific issues in the estimation of response surfaces are subsequently raised.

Typically, economists judge the appropriateness of a dose-response or production function model its desirable economic properties. Because of this penchant for economic properties, economists may overlook or ignore the biological or chemical processes that underlie the production relationships they are estimating [4]. Both conceptual and empirical validity of these functions may benefit from a fuller understanding of the production relationships than has traditionally been recognized by the economics profession.

For example, input and output variables in production theory are traditionally treated as flows or rates of use per unit time. These rates of use are viewed as proportional to some fixed stock of each input. The underlying stock and its conversion to output flows may be altered, however, if varying environmental states affect the optimal and actual output use rates over a given time interval. To visualize this, consider the following production relation [5]:

$$V = \int_{t_0+H}^{H+\mu} r(t)dt \tag{1}$$

where V = contemplated or actual volume of output
t_0 = the present
H = time at which the production activity is initiated
μ = time at which the production activity is terminated
r = rate of output

V, r, and μ determine the economic cost of this volume or level of output. The rates of flow and hence volume for different input/output combinations may be affected differentially by environmental factors working through basic chemical laws. The complex chemical linkages between inputs and outputs for many biological processes suggest that stock or capital theory considerations are of potential importance in assessing the true dose-response of a given ecosystem component.

Rather than imposing strictures on the form of the relationship (e.g., homogeneity) typically associated with functions of the Cobb-Douglas (CD) or constant elasticity of substitution (CES) type, Marsden et al. [4] show that modeling the response or production function to consider the underlying chemical processes can provide alternative and potentially useful functional specifications. While primarily concerned with engineering applications, their development does indicate that a more eclectic approach to functional form selection may be desirable for dealing with specific processes. Similarities between these relationships derived from chemical laws and the traditional CD or CES economic formulations suggest that a richer interpretation is possible if some understanding of the former is applied to the processes being modeled. While not explored further in this chapter, these dimensions of biological dose-response are worth noting in any analysis of processes based on chemical or biological laws, including environmental economic assessments.

Issues in Designing Studies of Response Surfaces

Anyone who proposes to engage in estimation of response surfaces must give pragmatic consideration to several practical and interrelated issues. All these issues require compromises with the abstract analytical frameworks of the applicable disciplines. A reasonably complete listing with particular relevance to the study of acid deposition-ecosystem component response surfaces might include the design of response surface experiments, estimation of these surfaces, choice of a model to represent the surface, and sources of discrepancies between response surfaces estimated in controlled or experimental conditions and observed in field conditions. We shall deal briefly with each of these issues in sequence, trying to highlight those features that seem particularly relevant to studies of the impact of acid deposition. A more complete treatment may be found in Adams and Crocker [6].

Experimental Design

The importance of biologically based information in economic assessments tends to increase as one moves from managed ecosystems associated with agriculture to more complex natural systems, because producer adaptations play a lesser role in the latter. For simple monocultures associated with agriculture, available secondary data on costs (inputs) and output or yields in the presence of producer behavioral assumptions may be used to infer the response, even the absence of any biologically based experiments. Such indirect procedures are also typically less expensive to perform than are direct biological experiments. For more complex natural systems, requisite economic data are not directly available. In this case, response information from biological experimentation may be the only feasible approach.

In situations where an experiment is the biologically appropriate way in which to generate and to test hypotheses about response surfaces, it is important that the economically relevant region of the surface be covered systematically. The great majority of biological research into response surface questions is of minimal use to the economist because it does no more than use analysis of variance techniques to establish only whether there exist statistically significant differences in the output obtained from a few levels of a single input. The traditional emphasis has been and continues to be on replication, where the replication is intended to improve precision. When the objective is to estimate a response surface, replication is much less essential. Primary concern should be with developing a model that predicts real-world outcomes better than the next best alternative. Experimental data should thus be generated so as to facilitate statistical comparisons of performance across alternative models, not just establish that the results of some *particular* model have statistically significant differences.

Changes in input mixes and magnitudes can substitute for replications of a particular input mix and magnitude, since both types of observations are intended to locate the response surface. For a given outlay of research resources, the information provided by more observations on output responses to an assortment of economically relevant input mixes and magnitudes will usually be more valuable than will the information garnered from additional replications using a particular input mix and magnitude. Moreover, if alternative models have similar *a priori* plausibility as descriptors of a response surface, empirical discrimination among models will obviously be assisted more by increasing the breadth and the density of the sampling coverage of the surface rather than by replication of experiments directed at only one point on the surface. A single or few points of the surface provide inadequate statistical and conceptual bases for judging model performance.

The potential conflict in the desires of biologists and economists with respect to the design of response surface experiments conducted with limited research resources can be minimized by examining criteria that may be used to weigh the tradeoff between replication and density of coverage. Anderson and Dillon [2] provide a detailed treatment of the efficiency conditions for this choice. Conlisk [7], Conlisk and Watts [8] and Morris [9] extend earlier treatments of optimal experimental designs to cases where the form of the response function is unknown and both the research budget and the number of experimental units are limited. In the absence of a specification of a particular design problem, the three universal implications of these conditions for response surface experimental design are rather simple and apparent. First, the greater the sensitivity of the system being investigated to variations in exogenous parameters, the greater the desirability of additional replication. Second, the greater the number of factors thought to impinge in nontrivial ways on system behavior, the more desirable is increased density and breadth of coverage of the economically relevant regions of the response surface. Third, since it is along these portions that outputs are sensitive to input mixes and magnitudes, research resources should be aimed at denser coverage and greater replication along the steeper parts of the economically relevant portions of the response surface.

In general, the essential fact of which the allocator of research resources must be aware is that there likely exist positive but declining marginal payoffs to additional observations drawn from any particular system or for *any* variable or particular combination of variables in that system thought to influence the response surface; that is, each additional observation adds something to the expected payoff, but these additions get progressively smaller as the number of observations increases. If the cost of research is a monotonically increasing function of the number of observations, one obtains the familiar optimality condition determined by the equation of marginal costs and marginal payoff.

Estimation of Response Surfaces

The appropriateness of available statistical methods for estimating response surfaces is well understood. Any good econometrics text, such as Kmenta [10], will provide a detailed and thorough treatment of the subtle issues of estimation that arise in a wide variety of commonly faced contexts, including joint outputs, nonlinearities in the parameters, observations which vary cross-sectionally and temporally, systems of equations, nonnormality of error terms across experiments on the same response surface, and truncated and censored dependent variables. Econometrics appears to have little to offer biometrics with respect to useful and correct applications of these techniques.

However, when the natural scientist uses field data rather than or along with experimental data to arrive at response surfaces, the perspective of the econometrician does have something valuable to offer. In particular, the econometrician will be sensitive to the implications for estimation of the fact that organisms behave "as if" they are making choices. Accurate estimation of the response surface parameters thus requires data on the factors that influence these choices. Moreover, an explicit representation of the organisms's choice problem must be built into the structure to be estimated. This choice paradigm is potentially as powerful a means of explaining the behavior of nonhuman organisms as it has been for human organisms. The importance of accounting for its influence even in a supposedly pure natural science exercise in estimating response surfaces is easily illustrated.

For example, assume the research problem to be the estimation, through a combination of field and experimental data, of the response of trout populations to acid deposition. (This illustration is an adaptation of a development in Crocker et al. [11].) In implicit form, a good approximation of the expression the natural scientist might apply to the field data collected over a given time interval is:

$$Y = f(X,W,Z,E,\epsilon) \tag{2}$$

where Y = stock of trout
   x = vector of aquatic ecosystem characteristics
   W = vector of weather characteristics during the period of analysis
   Z = measure of the fishing pressures imposed by humans on the trout stock
   E = measure of trout stock exposures to acid deposition
   ϵ = stochastic error term

The *a priori* information that experimental regimens have provided might be used to determine the functional form and the listing of variables on the right side of

Equation 2, to restrict the signs and/or the magnitudes of the coefficients of these variables, and/or to specify the properties of the error term. For simplicity, assume that Equation 2 is linear in the original variables. The coefficient attached to the acid deposition variable is then the reduction in trout stocks due to a one-unit increase in acid deposition. Would it then be reasonable to infer a dose-response association from the coefficient of this variable?

Such an inference would be correct if and only if it is possible to alter the acid deposition exposure without altering the value of any other explanatory variable in the expression. It is easy to show that this cannot be done unless the structure of the response surface is presumed to consist of no more than one relationship. More than one relationship is present in Equation 2; it contains a variable, Z, the levels of which have been and continue to be subject to control by fishermen. For example, the reduction in trout stocks due to exposures to acid deposition might be dependent on the number of mature fish capable of reproduction that fishermen have caught. To explain the trout stock outcome, the researcher must do more than simply enter the amount of fishing pressure: he must also explain the structure underlying the choice of the degree of fishing effort applied. One element in this choice will be the size of the trout stock. The following simple example shows one way in which trout stocks and fishing pressure might be jointly determined.

If both the acid deposition-trout stock response function and the fishing activity demand function can be linearly approximated, they can be written as:

$$Y = \alpha_1 + \alpha_2 E + \alpha_3 X + \alpha_4 Z + \alpha_5 W + \epsilon_1 \tag{3}$$
$$Z = \beta_1 + \beta_2 Y + \beta_3 I + \beta_4 P_N + \beta_5 P_Z + \epsilon_2 \tag{4}$$

Equation 3 states that the quantity of effort the fishermen choose to expend is related respectively to the trout stock, fishermen income, an index of the unit prices of substitute recreational activities and the unit price of fishing effort.

Solving Equations 3 and 4 for Y, we have

$$Y = \frac{\alpha_1 + \alpha_4\beta_1}{(1 - \alpha_4)\beta_2} + \frac{\alpha_2}{(1 - \alpha_4)\beta_2} E + \frac{\alpha_3}{(1 - \alpha_4)\beta_2} X + \frac{\alpha_4\beta_3}{1 - \alpha_4\beta_2} I$$
$$+ \frac{\alpha_5}{(1 - \alpha_4)\beta_2} W + \frac{\alpha_4 + \epsilon_2\epsilon_1}{(1 - \alpha_4)\beta_2} \tag{5}$$

Consider the coefficient attached to E in Equation 5. If E is acid deposition, Equation 5 shows that an estimate of Equation 3 will not yield the response of trout stocks to acid deposition, even though the dose-response function is "adjusted" for aquatic ecosystem characteristics, weather, and fishing effort. Instead, the coefficient for E in Equation 5 will be an amalgam of stock effects due to acid deposition, fishing effort, and the effects of trout stocks on fishing effort. The product of the coefficients for the latter two effects would have to approach zero for the response of trout stocks to acid deposition alone to be obtained. For this to occur, trout stocks could have no effect on the amount of fishing effort and/or fishing effort could have no effect on trout stocks. Both assertions are equally implausible. In fact, in the absence of further information, the sign that would be obtained for E when Equation 3 is estimated alone is ambiguous since $\alpha_2 \leqq 0$, $\alpha_4 \leqq 0$ and $\beta_2 \geqq 0$. It is entirely conceivable, if one were to estimate Equation 3 alone, that one would find that acid deposition enhances trout stocks. In any case, because the product of α_4 and β_2 is negative in sign, the effect of acid deposition on trout stocks will be underestimated. However, this negative bias in the response estimated is not predestined. Given Equation 7, a slightly different specification of Equation 6 could readily introduce a positive bias.

To attempt to account for the additional factors thought to influence an organism's response to acid deposition by simply stringing out variables in a single expression is often incorrect. During the period in which the response is supposed to occur, organisms can behave so as to influence the magnitudes assumed by certain of these variables. Each variable susceptible to this influence must be explained by an expression of its own if the purpose of the research is to explain the response of the organism to acid deposition rather than simply to predict its response. Unless circumstances are identical across space and time, predictions based on some version of Equation 8 will err for reasons no one will be able to identify until the response structure is comprehended. Because some human decision variables both influence and are influenced by the response, economic analysis is frequently necessary to impart an interpretable form to response expressions. Purely biological constructs will therefore often be insufficient tools with which to establish pollution response surfaces. They become even less sufficient if the units of analysis they employ deny the existence of substitution possibilities of interest to human and/or nonhuman decision-makers.

Choice of Models

The criteria for choosing among alternative models or theories of ecosystem behavior should relate to the value of information they provide. If two models have the same costs in terms of data requirements and application, the preferred model should be that which provides the greatest expected payoff. If the models differ in their costs, this difference should also be allowed for in the payoff appraisal. In general, the important question is not whether any particular type of model is biologically or statistically better than its alternatives, but whether it can better serve the objectives of decision-makers.

There is typically a tradeoff between model generality or elaboration and measurement error. Many researchers opt for increasing generality in model construction. Such generality usually requires complex models structured to encompass a large number of variables. The difficulty of empirically implementing such a model tends to offset quickly the benefits of generality. On the other hand, the ideal of many applied scientists is to design an experiment or research effort such that the scientist does not have to think about what the results mean: the answer the experiment gives is unequivocal. Attainment of this state requires that measurement be free from bias. The measurement errors that occur when this condition is not fulfilled can be reduced by devoting more resources to constructing measurement devices and techniques, by allowing more time for measurements to be made, and by better training of measurement personnel. However, measurement resources are expensive.

The tradeoff between (increased cost of) model elaboration and reduction in measurement error is addressed by Paratt [12]. As he shows, errors in model parameter estimation increase in the presence of correlation among explanatory variables. Generally, the greater the number of attributes introduced into a model in the form of properties that must be directly measured, the more likely are some pairs of these properties to be highly correlated. Relatively simple models, by definition, require fewer directly measured properties for their solution. Thus, simple models reduce the possibility of estimation biases.

In addition to the effects of correlation among explanatory variables, there are two more bases for evaluating the tradeoff between model complexity and errors in measurement. First, measurement resources are more likely to be allocated efficiently if they are assigned to those directly measurable properties thought to have a really significant influence on the derived property. Since the variables that have a significant influence on a derived property will frequently be the same in both complex and simple models, the use of the simple model is to be preferred if avoidance of substantial error in the estimate of the derived property is of high priority.

Second, it pays to devote resources to reducing the larger of these measurement errors, including those interactive properties whose products are greatest. Since in simple models there are fewer estimates of directly measured properties to be obtained, it follows that to a greater extent than in a complex model, a given stock of measurement resources can be used to reduce the error associated with any one property. Thus, given the cumulative nature of measurement error in models where measured properties are tied together in long chains of reasoning, this rule along with the previous two implies that simple models can be highly advantageous in estimating ecosystem responses to pollution. The advantages exist apart from the fact that simple models are relatively easy to use and, in spite of the interesting scientific details they may neglect, they will usually give quick answers to questions.

The preceding statements about the advantages of using simple models to describe response surfaces have not been made in the absence of empirical supporting evidence. For example Perrin [13], while studying the responses of various Brazilian crops to fertilizer applications, has contrasted the value to farmers of the information obtained from a simple structure based on Liebig's law of limiting factors (the law as succinctly stated by Swanson [14] says that yields increase at a constant rate with respect to applications of each factor unit if some other factor is limiting) to the information acquired from a multiinput, nonlinear (quadratic) representation commonly favored in much controlled fertilizer response research. Using a set of 28 experiments conducted at various Brazilian sites over a three-year period, he compared farmers' implied *ex post* net revenues from the two distinct models. If soil characteristics were accounted for, the simple one input, linear model based on Liebig performed equally as well as the nonlinear model.

Supporting evidence may also be gathered from an earlier effort by Havlicek and Seagraves [15], whose analysis of fertilizer response functions recorded little difference in economic consequences from using alternative model specifications. In this case, the economic value of information was not very sensitive to the form of the function, although standard statistical criteria would have suggested superiority of a cubic functional form.

Empirical evidence similar to Perrin [13] and Havlicek and Seagraves [15] is now beginning to appear for the connected black box simulation models so widely favored in much applied ecological research. Stehfest [16] has compared the payoffs from a simple Streeter–Phelps model of dissolved oxygen and a complex ecological optimal control simulation model with six state variables. Both models were built to provide information on the costs of meeting a water quality standard in a stretch of a West German river. The payoff was defined in terms of cost minimization. The total annual costs of meeting the standard when the water treatments suggested by a simple model were implemented were 8% lower than would have been the treatments recommended by the more complex model. Of course, the costs of establishing what constituted the recommended treatments were also lower for the simple model. Additional reviews of the performances relative to some objective of simple vs complex models are available [17–19]. Few, if any, implementable rules, other than those of Paratt [12] already remarked on, issued forth from these discussions. There is, however, general agreement that although it is naive to view simplicity per se as desirable, the research administrator should place the burden of proof that valuable information will be produced onto the proponents of proposals to build ever more complex ecological and economic models. Ultimately, the issue of the tradeoffs between simple and complex models can only be resolved empirically.

Whatever the virtues of model simplicity, it must be admitted that increases in model complexity are worthy attempts, in the absence of information acquisition costs, to improve model robustness, where robustness can be defined as the domain of circumstances where the model can be applied without undergoing structural revisions. However, as an alternative to the devotion of more and more research resources to molding, measuring and manipulating an ever-lengthening string of variables someone reasons or feels may influence what Young [19] terms a "badly defined system," axiomatic methods can be used. These methods permit inferences to be drawn about difficult-to-measure variables by deriving relationships between them and more readily observed variables. In addition, these axiomatic methods, before any attempt at measurement, allow discrimination between important and trivial contributors to system behavior. Suggestions for adoption of holistic methods [20] that recurrently appear in the biological literature are in the spirit of the axiomatic means of introducing information, as is the bioenergetics research of Bigelow et al. [21] and Hannon [22].

Experimental vs Field Response Surfaces

Generally, responses obtained under experimental conditions that set other, controlled variables at maximum yield levels will significantly exceed in absolute value the responses to be observed under field conditions. As a result, pollution control decisions based solely on experiment-derived response surfaces must be less than fully satisfactory. Indeed, these experimental results might best be viewed as untested hypotheses. Any generalizations drawn from such experiments concern input configurations not typically found beyond the experiment, in a set of exogenous parameters that nature may never replicate.

The reasons for discrepancies between experimentally derived and field observed responses surfaces are probably several. Two come readily to mind. First, as Anderson and Crocker [23] point out, so as to remove confounding sources of stress, all factors other than pollution that might influence behavior in controlled experiments tend to be set at biologically optimal levels. Given that these biologically optimal levels exceed those found in everyday environments, it follows that they are less binding, implying, by the Le Chatelier principle [24], that the contribution of a positive input (e.g., a pollution reduction) to the behavior parameter of interest will be greater than it otherwise would be.

A second, less obvious reason arises from the role that risk plays in managed ecosystems, particularly agricultural and forest systems. In strictly controlled experimental settings, all feasible sources of random variation in output levels are excised. However, in field conditions, the system manager must adapt his activities to natural sources of random variation such as weather, insect infestations and air pollution. As Just and Pope [25] demonstrate, the input mixes and magnitudes the system manager selects influence both the levels of output in any one time interval and the variability of these levels over time. Thus, for example, if the land area for which a farmer is responsible increases and he has no more inputs (e.g., lime, fertilizers and labor) than before, the susceptibility of his crops to a pervasive pollutant such as acid depositions will also increase. In taking countermeasures to such a pollution event, he has to spread the same inputs over a greater area. The implications of this as a source of discrepancies between experimentally derived and field observed response surfaces are developed by Adams and Crocker [6].

DECISION APPROACH TO THE VALUE OF RESPONSE INFORMATION

The previous section discussed key features of dose-response studies that will enhance the usefulness for the economist of the information they generate. This section presents a formalized framework for examining the value of such dose-response information in a policy setting. The decision problem is cast in terms of establishing an environmental regulatory policy. In this development, the main source of uncertainty in the decision problem is assumed to be the nature of the dose-response function. Hence, once the dose-response information is available, the economic consequences may be calculated. The decision process will also be assumed to be based solely on economic criteria; no other arguments are introduced into the decision-maker's objective function.

One reason (besides understanding fundamental biological mechanisms) for obtaining dose-response information on environmental pollutants is to set economically efficient policies for pollution control. For those charged with standard setting, numerous questions should be addressed concerning the acquisition of this dose-response information. For example, how much information is needed to set policy? How much precision is needed in the response function parameters? Once a level of information is obtained, what is the worth of additional observations (or precision) on the parameter in question? It is difficult to approach these informational issues via classical statistical procedures concerning type I and type II errors in hypothesis testing. That is, the emphasis in setting efficient environmental regulations may be to minimize the probability of type II error (falsely accepting the null hypothesis of no relationship between pollutants and environmental wellbeing) rather than the traditional concern with type I errors. Alternatively, Bayesian procedures as used in statistical decision theory can measure the *ex ante* value of sample information and hence provide a more meaningful response to the above questions (for an introductory reference, see Winkler [26]).

A hypothetical but plausible example of this type of decision problem may be that faced by EPA. Charged with shepherding a broad range of environmental parameters, EPA must not only monitor environmental quality but also must adopt policies intended to preserve "quality" at some designated level. In setting National Ambient Air Quality Standards (NAAQS), the Clean Air Act mandates that both health (primary) and welfare (secondary) effects of pollutants must be considered. While economic impacts are not considered in selecting concentrations for secondary standards, cost/benefit procedures measuring the economic consequences of pollution or the benefits of pollution control for a given standard are frequently employed. To arrive at these economic measures, there must frequently be some response information relating response of a particular entity or organism (say, an economically important crop) to the pollutant in question. Research aimed at obtaining this type of data would be funded by EPA's Office of Research and Development. If cost/benefit analyses (economic criteria) are to be used in the regulatory process, the research funds allocator needs to consider questions raised above regarding the worth of the information obtained and whether further funding of additional research will improve the economic measures and hence the policy performance of EPA.

The intent here is to present a simple but heuristically useful decision model of the above process. The problem will be structured as an *ex ante* analysis. That is, the *expected* value of information is assessed before the information actually becomes available. The *ex ante* structure would appear to be particularly relevant to the issue of acid deposition, where much dose-response information has yet to be acquired.

Decision Analysis

Some common links run through all decision analysis models. For example, decision problems have certain basic elements: actions, events, outcomes or consequences of the actions, probabilities of the events, and payoffs from the consequences or outcomes. The actions may be viewed as the alternatives available to a decision-maker. For the problem outlined above, these may include: (1) impose a pollution regulation or standard; or (2) to preserve the status quo (no regulation). The events may be the possible responses of the ecosystem to the pollution random variable (the actual level of pollution that occurs). The outcomes or payoffs are then functions of the benefits (to society) from the various actions taken and the probabilities of the events occurring, as objectively known or subjectively assigned by the decision-maker.

Given understanding of the behavior of affected economic agents, the dose-response function serves to determine the outcome or consequence of a given policy that, when combined with decision rules, generates the optimal course of action under the specified conditions. (No action may be viewed as an action in favor of the status quo.) The more information the decision-maker obtains, the less uncertain is the world. Thus, in this example, additional dose-response information reduces some of the variation and bias in prior estimates concerning how pollution impacts an ecosystem or ecosystem component.

These concepts may be formalized by introduction of appropriate notation and structural relationships. This is done below following the logic of decision analysis presented above and borrows from work by Katz et al. [27]. In the ensuing discussion, the major source of uncertainty will be the nature of the response; all other variables and factors will be assumed known.

Decision-Making Parameters

Decision-making parameters consist of the actions, events, consequences or payoffs and associated probabilities found in the decision problem. They are defined below in the context of the response information-regulatory problem. For simplicity, the decision-maker's utility function will be assumed linear over the relevant range, hence invoking the expected utility hypothesis will not change the ranking of events from that obtained using monetary values.

Actions

For ease of presentation, assume two possible actions: A_i, $i = 0,1$.

- Action A_0: set no standard affecting acid deposition (accept mean pollution level, μ); or
- Action A_1: set standard at level less than prevailing or mean pollution level, $\mu - \Delta$, $\Delta > 0$.

Alternative actions, each representing different (and perhaps more restrictive) pollution levels could be defined. Such an extension would render the specification more consistent with reality but would not add greatly to the heuristic emphasis of this development.

Setting a standard or policy more restrictive than that prevailing implies costs to those whose actions are being curtailed as well as to those who consume the polluting agent's output. Typical costs for air quality regulations are those associated with point source emission controls (e.g., scrubbers) and mobile sources such as automobile emission standards. Let costs $C_0 = 0$, and $C_1 =$ cost associated with action A_1.

Events

The events in this problem are the possible states of the pollutant or acid rain random variable X. Thus, if action A_0 is taken, then X has mean μ. If action A_1 is taken, then X has mean $\mu - \triangle$. Since X is a continuous random variable, an infinite level of pollutant concentrations is possible. Standard setting is a complex process and varies with the pollutant and perceived effects. As a result, standards imposed may not necessarily correspond to the average or mean concentration. Thus, some distribution (in this case log-normal) is assumed, from which probabilities of exceeding the standard may be inferred.

Consequences and Payoffs

Each action described under Actions has a corresponding consequence or outcome. In this example of agricultural damage due to acid rain, we need to obtain the mean yield for a crop (the conditional mean of Y) associated with a pollutant level $X = x$. This is provided by the pollutant dose-response functions, which are assumed to be linear:

$$E\ (Y|X = x) = \alpha - \beta x \tag{6}$$
$$\text{with Var}\ (Y|X = x) = \sigma^2 \tag{7}$$

In this analysis, the societal benefit produced by the agricultural system under various environmental parameters is measured in terms of the mean level of consumption realized at time t. Following the specification of Bradford and Kelejian [28], the societal benefit is given by the expected value of the Marshallian surplus W:

$$W = E \sum_{j=1}^{m} \left[\int_{o}^{Q_D} P_j^D(g_j)\ dg - \int_{o}^{Q_S} P_j^S(g_j)\ dg \right] \tag{8}$$

where   E = expectations operator
   $Q_D = Q_S =$ level of consumption at which quantity supplied and quantity demanded are equated
   $P_j^D(g_j) =$ demand price at time t for quantity g of the commodity j
   $P_j^S(g_j) =$ analogous supply price

For simplicity, we disregard intertemporal questions and transfer costs. To implement Equation 8 consider the following inverse linear demand and supply functions:

$$P_j^p = a - bg_j \qquad (9)$$
$$P_j^s = d + eg_j \qquad (10)$$

where   a,d = constants incorporating the effect of other variables in the demand and supply structure

Under such a structure, the integral defines W as a convex function of P.

The benefits measure reflects Marshallian surplus under a given set of economic and environmental conditions. Policies affecting the environmental conditions assumed to be subject to control are based on forecasts of the dose response. The value of dose-response information may then be expressed as $W_0 - W_1$ where W_0 is the value of Equation 8 under action A_0 (mean pollution level, μ) and W_1 is the corresponding value under action A_1 (pollution level $\mu - \triangle$).

The benefits measure defines the payoff or consequence from each course of action. Specifically, let $R(A_i)$, i = 0, 1, be the payoff when action A_i is taken, then

$$R(A_0) = W\ (Y|\mu) \qquad (11)$$
$$R(A_1) = W\ (Y|\mu - \Delta) - C \qquad (12)$$

Estimated Payoffs

In the absence of perfect information, decision-makers rely on estimates of true parameters or values. Because the true dose-response function is unknown, the payoffs must be estimated using a dose-response function estimated from a sample consisting of n pairs of observation $\{(X_k, Y_k): k = 1, 2, \ldots n\}$. The estimated dose-response function is of the form:

$$E\ (\hat{Y}|X = x) = \hat{\beta}x \qquad (16)$$

where $\hat{\beta}$ = estimate of the slope of the true dose-response function β

Typically, $\hat{\beta}$ is statistically obtained by the method of least squares:

$$\hat{\beta} = \frac{\displaystyle\sum_{k=1}^{n} X_k Y_k}{\displaystyle\sum_{k=1}^{n} X_k^2} \qquad (14)$$

where   n = sample size, critical to the precision associated with $\hat{\beta}$

The importance of n (sample size) is easily appreciated by reflecting on the role of sample size in the variance of the $\hat{\beta}$ estimate. If additional observations are costly to generate, then determination of "optimal" sample size would appear to be important to the regulatory process.

Given that the response parameter is an estimate, the payoffs then become estimated payoffs, $\hat{R}(A_i)$, $i = 0$, 1, or:

$$\hat{R}(A_0) = W(\hat{Y}|\mu) \tag{15}$$
$$\hat{R}(A_1) = W[\hat{Y}|(\mu - \Delta)] - C \tag{16}$$

Optimal Actions for Alternative Levels of Information

Decision-makers, by incurring acquistion costs, can acquire a nearly infinite range of information levels. For this example, three cases will be presented, two extreme or polar situations for comparison purposes and a middle or typical level of information implying moderate acquisition costs. These are discussed below.

Perfect Information

Decision-makers seldom, if ever, have perfect information. However, perfect information can provide a useful benchmark by which to judge the comparative efficiency of sample information. Under conditions of perfect information, the decision-maker will choose action A_i, $i = 0$, 1, that has the highest payoff $R(A_i)$.

The decision rule for the two action problem is:

- action A_0 is optimal if $W(Y|\Delta) < C$, and
- action A_1 is optimal if $W(Y|\Delta) > C$.

Hence,

$$\text{Max } R(A_i) = \begin{cases} W(Y|\mu), & \text{if } W(Y|\Delta) < C \\ W(Y|\mu - \Delta) - C, & \text{if } W(Y|\Delta) > C \end{cases} \tag{17}$$

Sample Information

Real-world decision problems consist of a decision-maker facing sample information, rather than perfect information. Here, the decision-maker will choose that action A_1 having the highest *estimated* payoff $\hat{R}(A_i)$.

$$\text{Max } R(A_1) = \begin{cases} W(\hat{Y}|\mu), & \text{if } W(\hat{Y}|\Delta) < C \\ W(\hat{Y}|\mu - \Delta), & \text{if } W(\hat{Y}|\Delta) > C \end{cases} \tag{18}$$

No Information

The polar case to perfect information is where the decision-maker has no information on the dose-response. This is the same as assuming $\hat{\beta} = 0$. Then the expected payoffs become:

$$\hat{R}(A_0) = 0 \qquad (19)$$
$$\hat{R}(A_1) = -C \qquad (20)$$

In this case, action A_0 is optimal. In other words, with no information available regarding the dose-response function, there should be no standard set, assuming that the decision-maker wishes to minimize the probability of failing to reject a false hypothesis ($\hat{\beta} = 0$).

Value of Information

The preceding discussion has dealt with the parameters of the decision problem and optimal decision rules. As currently structured, the decision problem may be solved using classical statistical procedures. The empirical analysis developed subsequently is based on the above problem. However, to value the estimated dose-response function before obtaining estimated parameters requires introduction of additional Bayesian decision concepts (e.g., Bayesian regression). Information is valued in an *ex ante* sense; that is, this development presents the expected value of information before the dose-response information is actually available. The expected value of information (EVI) is the most a decision-maker should be willing to pay for the information. Information here is assumed to be of value only if it changes a decision.

Value of Perfect Information (VPI)

The VPI is measured relative to having no information. Thus,

VPI = payoff with perfect information − payoff with no information =
$$\text{VPI} = \underset{i=0,1}{\text{Max }} R(A_i) - R(A_0) \qquad (20)$$

This leads to a VPI for each of two conditions:

$$0, \qquad \text{if } W(Y|\Delta) < C$$
$$W(Y|\Delta) - C, \quad \text{if } W(Y|\Delta) > C$$

Expected Value of Sample Information (EVSI)

Again, the point of reference will be the case of no information; the value of sample information is then measured relative to having no information. Thus, where ρ is a probability

$$\text{EVSI} = \rho(\text{select } A_0)R(A_0) + (1 - \rho)(\text{select } A_1)R(A_1) - R(A_0) \qquad (22)$$

or

$$\text{EVSI} = \rho[(W(\hat{Y}|\Delta) < C)]R(A_0) + (1 - \rho)[(W(\hat{Y}|\Delta) > C)]R(A_1) - R(A_0) \qquad (23)$$

The probability weighted payoffs reflect the different probabilities of adoption of the alternative policy actions given the distribution of β. The political process determines these probabilities.

Under the standard regression model assumptions,

$$\hat{\beta} \sim N\left(\beta, \frac{\sigma^2}{\sum\limits_{k=1}^{n} X_k^2}\right) \tag{24}$$

So

$$Z = \frac{\hat{\beta} - \beta}{\sigma\left(\sqrt{\sum\limits_{k=1}^{n} X_k}\right)^{-1}} \sim N(0,1) \tag{25}$$

$P(\text{select } A_1)$

$$= \rho\,[W\,(\hat{Y}|\Delta) > C] + (1 - \rho)\,[(\hat{Y}|\Delta) < \frac{C}{W}] = P\left[Z < \frac{(C/W) - (\hat{Y}|\Delta)}{\sigma\left(\sqrt{\sum\limits_{k=1}^{n} X_k^2}\right)^{-1}}\right] \tag{26}$$

$$= \Phi\left[-\frac{(C/W) + (\hat{Y}|\Delta)}{\sigma\left(\sqrt{\sum\limits_{k=1}^{n} X_k^2}\right)^{-1}}\right] \tag{27}$$

where $\Phi = N(0,1)$ is the distribution function

Given the above distribution function, the expected value of sample information can be calculated.

$$\text{EVSI} = \left\{1 - \Phi\left[\frac{(C/W) + (\hat{Y}|\Delta)}{\sigma\left(\sqrt{\sum\limits_{k=1}^{n} X_k^2}\right)^{-1}}\right]\right\} W\,(Y|\mu)$$

$$+ \Phi\left[\frac{(C/W) + (\hat{Y}|\Delta)}{\sigma\left(\sqrt{\sum\limits_{k=1}^{n} X_k^2}\right)^{-1}}\right] [W|Y\,(\mu - \Delta) - C]\,W\,(Y|\mu) \tag{28}$$

$$= \Phi\left[\frac{(C/W) + (\hat{Y}|\Delta)}{\sigma\left(\sqrt{\sum\limits_{k=1}^{n} K_k^2}\right)^{-1}}\right] [W\,(Y|\Delta) - C] \tag{29}$$

Comparison of EVSI and VPI

We have now shown that VPI may be expressed as a discontinuous relationship. That is,

$$VPI = \begin{cases} 0, \text{ if} & W(Y|\Delta) < C \\ W(Y|\Delta) - C, & \text{if} - W(Y|\Delta) > C \end{cases} \tag{30}$$

Similarly, the expected value of sample information on the dose-response function reflects the probabilistic aspects of the decision problem. Specifically,

$$EVSI = \Phi \left[\frac{(C/W) + (\hat{Y}|\Delta)}{\sigma \left(\sqrt{\sum_{k=1}^{n} X_k^2} \right)^{-1}} \right] [W(Y|\Delta) - C] \tag{31}$$

Finally, it is useful to compare the efficacy of EVSI with VPI. This provides an indication of the relative worth of the sample information. Thus:

$$\frac{EVPI}{VPI} = \left[\frac{(C/W) + (\hat{Y}|\Delta)}{\sigma \left(\sqrt{\sum_{k=1}^{n} X_k^2} \right)^{-1}} \right] \text{ if } W(Y|\Delta) > C \tag{32}$$

$$= 0, \text{ if } W(Y|\Delta) < C \tag{33}$$

This formalized development models a simple hypothetical decision problem. Equations 21–33 also suggest a framework for measuring the *ex ante* worth of dose-response information. Implementing the valuation procedure (calculation of EVSI) requires specification of parameter estimates for the major arguments in the model as well as invoking some specific assumptions concerning the distribution of the β or slope coefficient for the response function. This task is not performed explicitly in the analysis. The following empirical example, while *ex post* in nature, does suggest the potential utility of response information in the decision process.

DOSE-RESPONSE INFORMATION: AN EMPIRICAL EXAMPLE

This section presents tentative and ongoing results demonstrating both the use and value of some classes of dose-response information in the regulatory process. The empirical evidence is based on oxidant pollutant data. Attention is also directed at measurement of the value of (more precise) information concerning only one ecosystem production or output determining force (pollution), assuming other determinants of production are known. The heuristic example is expressed in terms of a managed ecosystem (agriculture), and this for only selected crops. These abstractions should not, however, detract from the general purpose of the analysis, which is to raise issues of concern to economists involving their use of biological data and to suggest a framework for assessing the relative utility of the information.

Agricultural Problem

Agriculture is sensitive to environmental changes. The effects of numerous pollutants on agricultural crop productivity has been and continues to be a popular research area. Some of the reasons for this research popularity are obvious:

1. Agriculture is a major sector in the U.S. economy with a 1979 farm value of output exceeding $100 billion [29]. Recent growth in farm exports has enhanced agriculture's role in the U.S. balance of payments. Its importance is even greater for selected regions.
2. The importance of food and fiber to the wellbeing of the population affords agriculture a special status at all political levels, as evidenced by federal and state cabinet-level agencies to deal with agricultural issues. This tends to ensure substantial research funds for agricultural research, as exemplified by Hatch funds to the land grant universities.
3. Because of agriculture's somewhat unique economic and political situation, substantial economic and physical data are monitored, collected and, hence, readily available.
4. Most agricultural crops are annuals grown under fairly controlled or managed conditions.

Thus, it is relatively easy to perform biological experiments on these types of plants vis á vis forests or other ecosystems with more complex time and input dimensions. Whether agriculture is the most fruitful area for ecosystem response research remains to be established. Some preliminary evidence suggests that potential agricultural economic damages due to pollution may be substantially less than other forms of pollutant damage [30]. However, because of the readily available economic and dose-response data bases for agriculture, it will serve as the empirical example of ecosystem response used here.

The pollutant of interest in this volume is acid deposition. As noted earlier, with the exception of one recent publication on soybeans [31], there is little in the way of acid deposition–agricultural crop response information. The remaining crop-oriented acid deposition studies found in the literature are typically derived from controlled laboratory experiments, which can at best serve to suggest hypotheses about potential crop response which are worthy of further research [32]. Preliminary acid deposition–crop response evidence shows effects ranging from positive (increased yields) to negative, depending in part on level of pH exposure and the crop in question. The state of knowledge concerning the impacts of acid deposition on agriculture thus appears unsettled, at least relative to acid deposition impacts on, say, aquatic systems [33].

This situation is in sharp contrast to current information on oxidants, where research on plant effects dates back at least three decades [34]. More recently, research attention has finally been devoted to biological effects of direct importance in economic assessment, namely changes in crop yields associated with ambient oxidant levels. Inquiries of this type are being directed by EPA's National Crop Loss Assessment Network (NCLAN), a coordinated research program whose objectives include an economic assessment of oxidant damage to major agricultural crops. Response information for selected crops drawn from NCLAN data will be used as a surrogate for acid deposition to implement the above valuation approach. Stated differently, the oxidant data can provide an *ex post* assessment of dose-response information for the regulatory problem defined earlier, given that response information is available and that regulatory policies have been enacted. Specifically, response data for corn used for grain, soybeans and cotton derived from NCLAN oxidant experiments are used in this example application.

The ozone dose-response functions for these three crops are presented in Table I [35,36]. A linear form consistent with those reported elsewhere [35] is used in the analysis. Response (yield) is measured in units of output (grams) and then converted to percent change for the assumed dose. Dose is measured as the seasonal mean 7-h-d^{-1} ozone concentration. The 7-h exposure period is from 9:00 a.m. to 4:00 p.m., the period in which stomatal activity and, hence, plant sensitivity to pollution is greatest.

These three crops serve as important indicators of economic damages to agriculture from air pollution, given their substantial contribution to total crop value in the United States. In 1979 the farm gate value of these commodities exceeded $37 billion, or about 60% of total crop output. Furthermore, regional concentration of corn and soybeans in the Midwest states and cotton in the South and Southwest suggests they may be subject to pollutant (acid deposition or ozone) damages. Markets for most agricultural commodities have been studied extensively, providing a rather extensive set of demand-and-supply relationships. Unfortunately, the existing demand estimates for these crops have certain limitations with respect

Table I.   Ozone Dose-Response Functions for Selected Crops

| | Variable | | | Percent Yield[b] Change from Ambient Levels[c] |
Crop	Intercept ($\hat{\alpha}$)	Ozone[a] ($\hat{\beta}$)	R^2	(%)
Corn (grain) [35]	247.8	260 (68)	0.65	4
Soybeans [35]	95.3	309 (109)	0.95	16
Cotton [36]	1095.6	3755 (352)	0.97	16

[a] Dose is the seasonal 7-h-d^{-1} mean concentration. Values in parentheses are standard errors.

[b] Yield is measured in grams per plant of harvested product.

[c] Yield change is the percent increase in yields associated with an improvement in air quality from ambient levels to background levels (0.025 ppm ozone).

to use in our analysis. The existing studies are designed for other purposes and employ differing time periods and different market structures. Further, our application of Marshallian surplus requires inverse demand functions, whereas the majority of commodity demand studies were estimated in the conventional quantity-dependent form. Thus, modification of the original equations to the inverse form could impart some degree of error to the measured Marshallian surplus.

As a result of these limitations, it was deemed more appropriate to estimate directly farm level demand-and-supply relationships rather than to modify previous estimates. A simple structure was assumed, abstracting from specific market outlets and some theoretically important causal factors. For example, the demand for the commodities is a derived demand—consumers do not consume corn, soybeans and cotton directly. Rather, livestock products and cotton goods consumption create the demand for intermediate products. Demand is assumed to be represented by the following structure:

$$P = f(Q, I) \tag{33}$$

where p = farm level price (actual)
        Q = quantity consumed
        I = per capita income (actual)

f. is linear. Actual dollar measures were chosen over deflated measures given the interest here on current measures of economic surplus. Since emphasis is on farm level price changes, abstracting from market outlets may be reasonable, assuming farm level price is determined more by total production than actions within individual markets. Income may also serve to pick up other excluded variables and can serve to reflect changes in international income developments. While perhaps plausible, the resultant estimates for this form must still be viewed as conditional on the nature of the assumptions we have employed.

For the supply functions, the following equally simple functional relationship was assumed:

$$Q_t = f(P_{t-1}, Pc_{t-1}) \tag{34}$$

where P_{t-1} = farm level price (actual) in period $t - 1$
        Q_t = quantity produced in period t
        PC_{t-1} = the price of a commodity that competes for the farmers resource commitment in period $t - 1$

The lagged structure conforms to the generally employed cobweb hypothesis: quantity produced in period t is a function of price in the preceding time period. Corn and soybeans are assumed to be the competing crop in each other's supply equation; for cotton, wheat was assumed to be the competing crop. The estimates from the above structure were inverted to arrive at the price-dependent estimates needed for the benefits function. Long-run equilibria are assumed for each commodity to negate the effect of the one period lag between supply and demand in the

Table II.   Estimated Demand and Supply Relationships

Crop	Adjusted[a] Intercept	Slope Coefficient	R^2	Crop Statistics (1979) [29]		
				Acreage (10^6 ac)	Production[b]	Value (10^6 \$)
Corn				70.9	7,763.8	18,569
Demand	5.95	−0.00048	0.71			
Supply	1.15	+0.00013	0.86			
Soybeans				70.5	2,267.7	14,039
Demand	13.25	−0.00333	0.87			
Supply	2.94	+0.00155	0.92			
Cotton				12.8	14.7	4,392
Demand	73.17	−0.00148	0.75			
Supply	39.11	+0.00107	0.74			
Total				154.2		37,000

[a] Independent variables, other than production, were evaluated at 1977–1979 levels and added to the intercept term.

[b] Production units are millions of bushels for corn and soybeans, and millions of bales (500 lb) for cotton.

calculation of economic surplus. A summary depicting magnitudes of the commodity supply and demand estimates and relevant crop statistics is presented in Table II.

The response data, when combined with data on the structure of the commodity markets in question, constitute important parameters in the information valuation framework outlined earlier. Another parameter requiring specification is the cost of alternative policies. Here, the level of resolution in the data further complicates the analysis. There is limited information on costs of environmental policies, at least in a form directly applicable to this exercise. For example, the cost of alternative oxidant air quality standards has been estimated to range from \$3 to \$9 billion per year, with the actual estimate a function of measurement procedures as well as the reference point or starting level of pollution [37]. Even with such estimates, however, it is difficult to trace these costs back to individual pollution control beneficiaries. Thus, what share of the \$9 billion in total costs of achieving an ambient air quality standard of 0.08 ppm should be charged against agriculture, or against materials, etc.? One procedure (that employed here) is to prorate costs based on relative benefit. It has been suggested that total agricultural benefits from improved air quality may not exceed approximately 10–20% of total air quality benefits [30]. While supported by limited empirical evidence, we employ the 10–20% solution here to arrive at cost estimates. Thus, if the cost of improving air quality by, say, 20% is \$3 billion, then agriculture's share may be assigned as \$300–600 million. While admittedly a gross approximation at best, partitioning of benefits and costs in this fashion is an uncomfortable necessity until alternative measures are available.

Numerical Example

The following numerical example will first trace out the value of information in a regulatory context by comparing a set of progressively more stringent regulatory policies: A_0 which corresponds to the "no standard" case to action A_3, which represents a very restrictive oxidant standard. The next, and perhaps more meaningful, numerical analysis will estimate the worth of more precise dose-response information. Thus, once an estimated response coefficient ($\hat{\beta}$) is available, the question becomes how important or valuable are *additional* observations on the response in question in terms of making economic assessments; that is, how sensitive are the economic consequences to the variability in $\hat{\beta}$?

Four specific oxidant air quality standards are examined. It should be emphasized that the standard refers to peak hourly levels, not annual average levels. Thus, if air pollution is assumed to be log-normally distributed, a standard of 0.12 ppm (not to be exceeded more than once per year) translates into an average ambient concentration of approximately 0.06 ppm annual average. Action A_0 corresponds to a "no standard" case, which we assume here to be an (arbitrary) standard of 0.18 ppm (or an annual average of 0.09 ppm). Action A_1 portrays the current national ambient air quality standard of 0.12 ppm not to be exceeded more than once per year. Actions A_2 and A_3 are progressively more restrictive standards of 0.10 and 0.08 ppm, respectively.

By using the specific crop dose-response information from Table I and the noncompensated farm level commodity supply and demand relationships from Table II, the economic surplus for each action may be estimated. Specifically, this is done by integration of the benefits function (Equation 11) at relevant equilibria, as determined by pollutant induced supply shifts. The A_1 or current production level for each commodity is portrayed by the average production level over the 1977–1979 period. The results of the integration for each action, along with shifts in surpluses across actions, are presented in Table III.

A number of observations may be gleaned from the table. For example, while the economic surplus measures increase with improvement in air quality, the impact varies across crops. Such variation is a function of the underlying economic and biological parameters and indicates the importance of broad crop coverage. The magnitudes of the surplus measures (~$25–30 billion) seem plausible; differences in surpluses indicate a rather sizable benefit (~$3.4 billion) is accruing to the current standard vis á vis no standard. The benefits decline, however, as more stringent policies are imposed (e.g., A_2 and A_3).

The present oxidant standard, as measured by A_1, appears to be yielding a substantial improvement in consumer wellbeing. However, to place these values in the true context of the worth of the policy or standard, the costs associated with each action must also be recognized. This is done in Table IV, where the costs are the prorated share of total compliance costs as assigned to agriculture. Again, the net benefits of the present standard appear to be in excess of the apportioned costs. Assuming the present standard to be mostly the result of extant dose-response information, the value of such information as manifested in the

Table III. Marshallian Economic Surpluses in Billions of 1979 Dollars for Alternative Oxidant Policy Actions

Crop	Actions[a]				Change in Surplus		
	A_0	A_1	A_2	A_3	A_0 to A_1	A_1 to A_2	A_2 to A_3
Corn (for grain)	14.323	15.297	15.653	15.941	0.974	0.356	0.288
Soybeans	7.210	9.325	10.002	10.681	2.115	0.677	0.679
Cotton	0.855	1.133	1.218	1.313	0.278	0.085	0.095
Total	22.388	25.755	26.873	27.935	3.367	1.118	1.062

[a] Action A_0 is the case of no oxidant standard. In this analysis, "no standard" is assumed to equal an ambient oxidant standard of 0.18 ppm. Action A_1 corresponds to the current national ambient air quality standard of 0.12 ppm. Action A_2 portrays an improvement in air quality associated with an air quality standard of 0.10 ppm. Action A_3 reflects a more restrictive standard of 0.08 ppm.

Table IV. Alternative Regulatory Actions and Associated Payoffs in Billions of Dollars

Action	Benefits (change in surplus)	Costs [37][a]	Payoffs[b]
A$_1$ (Impose Standard of 0.12 ppm)	3.367	0.600–1.200[c]	2.167–2.767
A$_2$ (Impose Standard of 0.10 ppm)	1.118	0.100–0.200[c]	0.918–1.018
A$_3$ (Impose Standard of 0.08 ppm)	1.062	0.180–0.360[c]	0.702–0.882

[a] Costs, as reported here, are a prorated portion of the *total* costs associated with the standard in question. Costs are expressed as 1978 dollars.

[b] Payoffs provide a crude upper bound measure of the value of the response information. Since the decision is a manifestation of other considerations in addition to direct effects on crops, the payoff accrues to all information used in the decision process.

current standard would be quite high. In reality, the policy currently in effect probably reflects multiple sources of information or input that constitute or affect the political calculus. This "value of information" then does not accrue entirely to the dose-response function and may best be viewed as the value of the standard or an upper bound on the response information. Finally, note that alternative policy actions imposing more restrictive standards could also be potentially beneficial, as reflected in the positive benefits reported to actions A$_2$ and A$_3$.

That some information is of potentially high value in setting regulatory policy is not a particularly novel or revealing observation. In this example, the dose-response information is assumed to derive from certain experimental procedures. Perhaps other, less traditional, information would also have provided acceptable policy input. Even without formalized experimental data, it is unlikely that the decision-maker would be completely without information. Hence, the no-standard case, which assumes no information, is perhaps an unreasonable benchmark or reference point, and results in an overstatement of benefits to the alternative standards.

The more interesting and meaningful assessment concerns the worth of precision in the response estimate, where precision is assumed to be obtained from additional observations or information on the same response. The values generated in Tables III and IV are derived from statistical estimates ($\hat{\beta}$) of a true but unknown response parameter (β). Each of these estimates has an associated variance. Increasing the sample size (additional observations) should reduce the variability around the estimate, ultimately converging on some stable value of $\hat{\beta}$ that approaches β. The worth of precision in this exercise pertains only to any changes in policies that may result from new, more precise estimates of β, i.e., precision is not viewed as having intrinsic value. Given that regulatory performance here is measured only in economic terms, the issue centers around the sensitivity of the economic estimates to the $\hat{\beta}$ values. This issue is explored by determining the sensitivity of the surplus measures to the variability in the response estimates, as measured by the standard errors around the regression coefficient estimates. The variability in the supply and demand parameters, which may be substantial, is not addressed here. Only the variability in physical response enters the benefits calculation. The results of the sensitivity analysis are reported in Table V.

As is evident from Table V, the variability around the predicted yield response is much greater than that around the estimated economic consequences of the response. For example, the coefficient of variation for the response ranges up to 29%, whereas the economic surplus 95% confidence interval (as measured by two standard deviations around the estimate for this sample size) is only 4%. The low variability in the economic benefits calculation reflects both the low absolute yield effect of ozone on some crops (e.g., corn) and the interplay of supply and demand estimates (slopes) from the economic models. Whatever the forces behind this observation, the implication is that, at least in terms of economic measurements, the variability in physical response may not be of great consequence in standard setting. When viewed in the regulatory context, the variation in the surplus measure would not affect the optimal course of action, i.e., the decision-maker is not likely to accept falsely the hypothesis of $\hat{\beta} = 0$ and impose action A_0, say, instead of A_1 under the distribution of $\hat{\beta}$ and associated benefits for this analysis. The implication of this observation seems quite important: there appears to be little support for continuous fine tuning of research on an effect or particular response, at least if economic consequences of discrete oxidant standards are to be used as the criterion in policy setting. This also suggests broader, less detailed response coverage than biologists and physical scientists are wont to employ. The allocator of public research resources must weigh these considerations, but if response research is primarily directed at economic consequences of alternative policies, then continued replication of publicly supported experiments in the name of traditional scientific procedures may not be the most efficient use of research resources.

Table V. Sensitivity of Economic Surplus Measures to Variation in Response Estimates (β)

Crop	Predicted Yield Response: Coefficient of Variation[a]	95% Confidence Interval: Economic Surplus for Action A_1 ($ billions)	Confidence Interval as Percentage of Mean
Corn (for grain)	26	15.653 ± 0.143	1.8
Soybeans	29	10.002 ± 0.382	7.6
Cotton	12	1.218 ± 0.058	4.8
Total		26.873 ± 0.583	4.4

[a] Coefficient of variation is the standard deviation of the predicted yield response divided by the predicted yield value.

CONCLUSIONS

Response information is very often an integral part of economic assessments aimed at pollutant damages. Growing interest in effects of acidic depositions on ecosystems and the demand for possible regulatory actions is generating additional response research. This chapter has presented some general economic issues concerning response estimation, including its role in economic assessments. It has also been demonstrated, in an *ex post* function, how one may judge the utility of response information in a decision-making environment.

The intent has been primarily heuristic: to demonstrate, in theory and with application, some economic issues concerning the generation and use of response information. The numerical example, while plausible, must be recognized as highly conditional. Any similar empirical assessment of response estimates would benefit from more comprehensive treatment of the economics of commodity demand and supply and better cost data, as well as a more complete statement of the decision-making process of regulatory entities. The assumptions are based only on economic efficiency arguments; practical decision-making and environmental regulatory policy setting are more complex than that.

DISCLAIMER

Although the research described in this article has been funded fully or in part by the U.S. Environmental Protection Agency through a grant to the University of Wyoming, it has not been subjected to the agency's required peer and policy review and, therefore, does not necessarily reflect the views of the agency, and no official endorsement should be inferred.

REFERENCES

1. Heady, E.O., and J.L. Dillon. *Agricultural Production Functions* (Ames, IA: Iowa State University Press, 1960).
2. Anderson, J.R., and J.L. Dillon. "Economic Considerations in Response Research," *Am. J. Agric. Econ.* 50:130–142 (1968).
3. Henderson, J.M., and R.E. Quandt. *Microeconomic Theory: A Mathematical Approach* (New York: McGraw-Hill Book Company, Inc. 1971).
4. Marsden, J., D. Pingry and A. Whinston. "Engineering Foundations of Production Functions," *J. Econ. Theory* 9:124–140 (1974).
5. Alchian, A. "Costs and Outputs," in *The Allocation of Economic Resources,* M. Abramovitz, Ed. (Stanford, CA: Stanford University Press, 1959).
6. Adams, R.M., and T.D. Crocker. "Dose-Response Information and Environmental Damage Assessments: An Economic Perspective," *J. Air Poll. Control Assoc.* 32:1062–1067 (1982).
7. Conlisk, J. "Choice of Response Functional Form in Designing Subsidy Experiments," *Econometrica* 41:643–655 (1973).

8. Conlisk, J., and H. Watts. "A Model for Optimizing Experimental Designs for Estimating Response Surfaces," *J. Econometrics* 11:27–42 (1979).

9. Morris, C. "A Finite Selection Model for Experimental Design of the Health Insurance Study," *J. Econometrics* 11:43–61 (1979).

10. Kmenta, J. *Elements of Econometrics* (New York: MacMillan Publishing Company, 1971).

11. Crocker, T.D., W. Schulze, S. Ben-David and A.V. Kneese. "Experiments in the Economics of Air Pollution Epidemiology," EPA 600/5–79–001a, U.S. EPA Washington, DC (1979).

12. Parratt, J.G. *Probability and Experimental Errors in Science* (New York: John Wiley & Sons, Inc., 1961).

13. Perrin, R.K. "The Value of Information and the Value of Theoretical Models in Crop Response Research," *Am. J. Agric. Econ.* 58:54–61 (1976).

14. Swanson, E.R. "The Static Theory of the Firm and Three Laws of Plant Growth," *Soil Sci.* 95:338–343 (1963).

15. Havlicek, J., Jr., and J.A. Seagraves. "The Cost of the Wrong Decisions as a Guide in Production Research," *J. Farm Econ.* 42:157–168 (1962).

16. Stehfest, H. "On the Monetary Value of an Ecological River Quality Model," Research Report RR-78-1 International Institute for Applied Systems Analysis, Laxenburg, Austria (1978).

17. Beck, M.B. "Some Observations on Water Quality Modeling and Simulation," in *Modeling Identification and Control in Environmental Systems,* G.C. Vansteenkiste, Ed. (Amsterdam: North-Holland Publishing Company, 1978), pp. 775–786.

18. Griliches, Z. "Estimating the Returns to Schooling: Some Econometric Problems," *Econometrica* 45:121 (1977).

19. Young, P. "General Theory of Modeling for Badly Defined Systems," in *Modeling, Identification, and Control in Environmental Systems,* G.C. Vansteenkiste, Ed. (Amsterdam: North-Holland Publishing Company, 1978), pp. 103–135.

20. Jorgensen, S.E., and H. Mejer. "A Holistic Approach to Ecological Modeling," *Ecol. Model.* 7:169–189 (1979).

21. Bigelow, J.H., C. Dziter and J.C.H. Peters. "Protecting an Estuary from Floods—A Policy Analysis of the Oosterschelde, Volume III, R-121/3 Neth," The Rand Corporation, Santa Monica, CA (1977).

22. Hannon, R. "Total Energy Costs in Ecosystems," *J. Theoret. Biol.* 80:271–293 (1979).

23. Anderson, R.J., Jr., and T.D. Crocker. "The Economics of Air Pollution: A Literature Assessment," in *Air Pollution and the Social Sciences,* P.B. Downing, Ed. (New York: Praeger Publishers, 1971), pp. 133–165.

24. Silberberg, E. *The Structure of Economics* (New York: McGraw-Hill Book Company, Inc., 1978).

25. Just, R.E., and R.D. Pope. "Production Function Estimation and Related Risk Considerations," *Am. J. Agric. Econ.* 61:276–284 (1979).

26. Winkler, R.L. *An Introduction to Bayesian Inference and Decision* (New York: Holt, Rinehart, and Winston, Inc., 1972).

27. Katz, R.W., A.H. Murphy and R.L. Winkler. "Assessing the Value of Frost Forecasts to Orchardists: A Dynamic Decision-Making Approach," *J Appl. Meteorol.* 21:72–85 (1982).

28. Bradford, D.F., and H.H. Kelejian. "The Value of Information for Crop Forecasting with Bayesian Speculators: Theory and Empirical Results," *Bell J. Econ.* (Spring 1978), pp. 123–144.

29. "Agricultural Statistics, 1980," U.S. Department of Agriculture, Washington, DC (1981).

30. Crocker, T.D. "Pollution Induced Damages to Managed Ecosystems: On Making Economic Assessments," in *Effects of Air Pollution on Farm Commodities,* J.S. Jacobson and A.A. Miller, Eds. (Arlington, VA: Issak Walton League, 1982), pp. 103–124.

31. Evans, L.S., K.V. Lewin, C.A. Conway and M.J. Patti. "Seed Yield (Quantity and Quality) of Field-Grown Soybeans Exposed to Simulated Sulfuric Acid," *New Phytologist* 89:459–470 (1981).

32. Lee, J.J., G.E. Neely, S.C. Perrigan and L.C. Grothaus. "Effect of Simulated Acid Rain on Yield, Growth, and Foliar Injury of Several Crops," *Environ. Exp. Bot.* 21:171–185 (1981).

33. "Proceedings of the First International Symposium on Acid Precipitation and the Forest Ecosystems," USDA Forest Service Technical Report NE-23, Upper Darby, PA (1976).

34. Middleton, J.T., J.B. Kendrick, Jr. and H.W. Schwolm. "Injury to Herbaceous Plants by Smog or Air Pollution," *Plant Dis. Rep.* 34:245–252 (1950).

35. Heck, W.W., O.C. Taylor, R. Adams, G. Bingham, J. Miller, E. Preston and L. Weinstein. "Assessment of Crop Loss from Ozone," *J. Air Poll. Control Assoc.* 32:352–361 (1982).

36. Taylor, O.C. "NCLAN Cotton-Ozone Response Information for 1981, Shafter, California," unpublished information (1982).

37. "Costs and Economic Impact Assessment for Alternative Levels of the National Ambient Air Quality Standards for Ozone," EPA 405/5-79-002, Office of Air Quality Planning and Standards, U.S. EPA, Research Triangle Park, NC (1979).

CHAPTER 5

Scientific Truths and Policy Truths in Acid Deposition Research

Thomas D. Crocker

Natural science literature scenarios on the sources and the effects of acid deposition are heavily larded with "cans" and "coulds." Uncertainty about cause-effect relations and even about the actual existence of some hypothesized effects is the dominant impression the careful reader obtains. As do the common local air pollutants, acid deposition stands accused of posing threats to existing stocks of organisms and artifacts [1–3]. Moreover, acid deposition is frequently indicted for mining nutrients and accumulating toxics in the earth's crust and its freshwater bodies [4,5], while simultaneously generating episodic damage events [6], reducing on regional scales the biological diversity of aquatic and terrestrial ecosystems [7,8], and exhibiting declining incremental damages over substantial intervals of increasing pollution concentrations [9,10]. In North America, these effects, which are thought to occur most frequently in the continent's northeastern portion, are often said to be caused primarily by precursor emissions of midwestern origin [11]. This assignment of responsibility has been reinforced by displays [12] depicting a coincidence of the supposed gradual spread of acid deposition from its 1950s focus in the Ohio River Basin with a near-doubling since then of eastern North American sulfur dioxide emissions and a threefold increase in nitrogen oxide emissions.

Literature which disputes the above precursor emission/long distance transport/deposition/effects story is easily found. Both accuracy [13] and precision [14] are said to be lacking in the instrumentation used to measure acid deposition. Evidence has recently emerged [15] which supports contentions that the emissions causing acid deposition in Rhode Island are primarily of local origin. Conclusions that the historical trend of acid deposition has been strongly upward in the last several decades have been countered by somewhat believable arguments that the data fail to uphold these judgments [16]. Any upward trend that does exist is said to have been greatly exacerbated by a federal regulatory emphasis on engineering control and dispersal devices such as scrubbers, precipitators and tall stacks [17]. Numerous experimental studies purport to demonstrate that acid deposition occasionally fertilizes agricultural [18] and forest corps [19]. Even though the aquatic impacts of acid deposition are regarded as reasonably well defined [20], their economic consequences are blurred. Just as acid deposition may improve the nutrient status of some soils, it may enhance the recreational stature of some freshwater bodies. Water clarity can be increased, fish communities may be rid of some undesirable species, and remaining fish may be larger. Finally, the geographical scope and the intensity of all these plausible positive and negative terrestrial and aquatic impacts is clouded: the assigned areal and temporal distributions vary wildly across the published studies [21,22], mainly because there is little prior agreement on the criteria to use to construct sensitivity indices.

Policymakers viewing the above evidence might conclude that former Senator Muskie's famous two-armed scientist ("On the one hand . . . , but on the other hand . . . .") rules discussions of the benefits of controlling acid deposition. Interested parties of one or another persuasion naturally try to exploit this even-handedness. Several of the papers in Cole [23] provide rather stark, side-by-side examples of this. Opponents of control, those to whom the expected net benefits of control are negative, insist that more information is required before "rational" control decisions can be made. They would opt for the status quo, or even a loosening of the bindings in which extant controls place them. On the other hand, proponents, those to whom the expected net benefits of control are positive, fear the possible environmental effects of delay—they want more controls now. Proponents emphasize the costs of delay in terms of the known and suspected environmental damages that immediate emission controls will prevent. Opponents point to the costs that will remain unborne if no controls are adopted. Proponents and opponents each possess unique standards of truth on which they would have policymakers base control decisions. That which is "conservative" and "scientific" to the opponent of control is flagrantly biased to the control proponent. I shall discuss some of the implications for acid deposition precursor emissions and emission impacts of alternative policymaker treatments of these two different standards, assuming a once-and-for-all control decision must be made. This somewhat discursive exercise is intended to illustrate an economic framework within which applications of the two truth standards might be evaluated rather than to provide an exhaustive analysis. Page [24], in a little-noticed but most insightful paper, has outlined, in an environmental policy context, the tradeoff between the two truth standards. Jacobson [25] has, in effect, raised the same issue for acid deposition. Bazelon [26] and Ricci and Molton [27] speak to the approaches that the courts and administrative agencies have adopted when confronted with this tradeoff. This last paper motivated the present effort. All these sources recognize that the owner of an inventive mind having the patience to engage in fishing expeditions can always find a plausible exception to any statement presented as unvarnished truth.

TYPE I AND TYPE II ERRORS

Opponents of control (emitters) would base acid deposition policy decisions on a scientific or cigarette company standard of truth; that is, they wish to minimize the probability of the policymaker treating as true some hypothesized environmental impact or impact source that really is not true. Just as the American Tobacco Institute continues to insist that the etiology of cigarette smoking and lung cancer has not been "scientifically demonstrated," the more radical of emitters might accept nothing less than that evidence of environmental impacts obtained from controlled experiments conducted under the most rigid of modern design protocols. Conversely, the more extreme of control proponents (receptors) would found policy decisions for control on the merest of associations between precursor emissions and possible environmental impacts; that is, they desire to minimize the probability of the policymaker treating as false some hypothesized environmental impact that really is true. Each would view the outcome of a hypothesis test as another bit of evidence to be set before some arbitrator or judge. Any decision the policymaker makes compromising these polar standards will be viewed as wrong; the policymaker might then be considered to have made the right decision when he has minimized the cost of making the wrong decisions. He must be predominantly concerned with how to gamble, not what to believe. As Olson [28] has remarked, there is no such thing as a safe lunch when dealing with regional environmental problems like acid deposition. The design of control policy is thus locked into a world of abstractions because we are unable to conduct the trials necessary to refine and to test our understanding.

In the literature of statistics, that which the emitter desires to minimize is known as type I error—the chance of a false positive, or the risk of treating emissions as causative negative agents when they are in fact benign. The receptor's objective is to minimize type II error—the chance of a false negative, or the risk of treating emissions as benign when they are in fact causative negative agents. Obviously the risk of perceiving benign agents as causes can be reduced by automatically treating almost all agents as benign. This can be done, however, only by increasing the risk of disregarding some truly guilty agents. Similarly, the risk of disregarding a guilty agent can be reduced by requiring less stringent evidence for assigning causes, but this would increase the risk of blaming a truly innocent agent for negative environmental impacts. In principle, the only way both risks can be simultaneously reduced is by acquiring more information about the structure of the underlying system and/or by obtaining more observations on the behavior of the system. The two kinds of risks are inversely related.

This inverse relation introduces a tradeoff that is readily translated into the language of expected values and is consistent with Neyman and Pearson's [29] original conception of hypothesis testing as the art of "gambling with truth" [30] rather than the art of weighing evidence. For example, if acid deposition precursor emissions are perceived initially to be a cause of ecosystem stresses, and the policymaker therefore decides to tighten emission controls, he incurs a risk of type I (false positive) error with its associated unnecessary increased control costs, v. One component of the expected cost of any policymaker decision is then Pv, where P is the probability of *no* ecosystem stresses in the absence of tightened controls. This expected cost is part of the policymaker's gamble, since subsequently acquired information may demonstrate that, retrospectively, these increased control costs could, at no cost to the environment, have been avoided. When, however, the policymaker initially views emissions as innocent and thus opts for no further controls, he risks a type II (false negative) error and the pecuniary equivalent of its associated ecosystem stresses q, which are yet another component of his gamble. Since the alternative hypotheses are exhaustive and mutually exclusive, the expected cost of a policymaker decision not to increase control is then $(1 - P)q$, where $(1 - P)$ is the probability of ecosystem stresses in the absence of more control. If, then, the policymaker tries to minimize the economic burden of foregone opportunities, he must minimize $Pv + (1 - P)q$. For a detailed treatment, see Winkler [31]. When they fail to indicate an environmental impact, research procedures designed to minimize only the probability of type I error say nothing in particular about whether or not an impact will occur. There are not many persons who will blissfully ride in an airplane when its chances of crashing are 90 out of 100, but will adamantly refuse to ride in this same plane when these chances reach 95 out of 100.

To consider this minimization problem more carefully, I exploit a framework initially set forth by Erlich and Becker [32] and Brown [33]. Resistance and emission control activities produce states of knowledge about pollution cause-effect linkages. Assume that the policymaker, the identical emitters, and the identical receptors are risk neutral. Receptors do not engage in strategic behavior with respect to each other, nor do emitters. Let x be a measure of a target resistance of the environment to acid deposition. This target resistance can originate partly in the efforts (e.g., the liming of freshwater bodies, or the application of zinc coatings to steel) receptors could make to protect their assets. During the time interval in question, let y be a measure of the target level of emission controls. Given that the policymaker contemplates actually reaching these targets during the period in question, the function $P(x,y)$ is the chance he will make a type I error; it depends on x and y such that $P_x > 0$, $P_{xx} < 0$, $P_y > 0$, $P_{yy} < 0$, and $P_{xx}P_{yy} - (P_{xy})^2 > 0$, where the subscripts indicate partial derivatives. A more resistant environment is more likely to give off false stress signals; thus $P_x > 0$. However, the power of investigatory techniques to distinguish between placebo and true stress signals seems likely to diminish with increasing resistance to stress; thus $P_{xx} < 0$. As for $P_y > 0$ and $P_{yy} < 0$, greater emission controls require that proponents of further control exercise their imaginations to shift public and policymaker attention to the more subtle and delayed possible kinds of environmental impacts. The controls will have washed away the more obvious kinds of impacts and the set of environmental components seen as being potentially susceptible must be expanded, though, given a finite set of components, it becomes progressively harder for quixotic types to discover new candidates. The last assumption about signs, $P_{xx}P_{yy} - (P_{xy})^2 > 0$, guarantees an interior solution to the minimization problem. It says that the effects on type I error from a change, if any, in resistance caused by a change in emission controls (or vice versa) cannot dominate the direct type I error effects of a change in either resistance or in emission controls. These assumptions on the signs of the $P(x,y)$ derivatives may be debatable, however they seem to be intuitively more reasonable than the alternatives.

Let the marginal cost of achieving a unit more resistance be c, and the marginal cost of controlling a unit more precursor emissions be w. Both, for simplicity, are assumed constant. The total cost associated with any particular prospective level of ecosystem protection via receptor defensive moves and adaptations and emitter controls is then $cx + wy$. If the policymaker adopts more environmental protection, this will be the cost of his making a type I error. Given that he does choose to adopt more protection, these costs are nevertheless borne whether or not his decision ultimately proves to be correct or incorrect, i.e., whether or not he has made a type I error. The socially optimal values of x and y are those that minimize social costs, $S(x,y)$; that is:

$$\min_{x,y} S(x,y) = cx + wy + D[1 - P(x,y)] \tag{1}$$

where, when further environmental protection is not adopted, D is the sum of additional environmental damage costs and additional control costs (from retrofitting) given that, contrary to the bets that had been placed, environment damages are realized. For simplicity, D is treated as independent of x and y. At least on the damage side, this is not altogether unrealistic for, as Crocker and Forster [34] note, many ecosystems such as freshwater fisheries exhibit sharply declining incremental damages from acid depositions. Small quantities of depositions are thus responsible for the major share of damages once threshold acidification levels are attained.

The entire last term in Equation 1 is the expected cost of a type II error, or the expected costs of having to retrofit emission controls and to bear environmental stresses that occur because acid deposition had, with perfect hindsight, been wrongly treated as innocent. Environmental stress resistances x and precursor emission controls y are therefore allowed to impact the *expected* value of D.

Minimization of (1) requires that:

$$c = DP_x \tag{2}$$

and

$$w = DP_y \tag{3}$$

The cost-minimizing solution, $S(x^*,y^*)$, to Equation 1 is found by solving Equations 2 and 3 simultaneously for x and y. Solution values, x^* and y^*, are those the policymaker would pick if he could select x and y and if he were serious about minimizing the costs of wrong decisions. Alternatively, they can be viewed as the levels of resistance and emission control a self-polluter would select who lacked full knowledge about the impacts of his emissions on his environmental assets.

In the acid deposition case, of course, there is neither a single, all-powerful policymaker able to dictate x^* and y^* at will, nor is there a single self-polluting individual who can be left alone to choose x^* and y^* in his own best interest. Instead, there are numerous owners of emitting and receiving assets who generally live in different regions, and who have only weak, if any, incentives to treat the costs they impose on each other as their own. The policymaker's tasks are to evolve criteria for weighing alternative truth standards, and to develop burden-sharing rules that are likely to cause the parties to generate information which simultaneously reduces the likelihood of type I and type II errors while causing voluntary choices of x* and y*. Except in the absence of inescapable conditions, fulfillment of these tasks is near impossible. One might seek a cooperative equilibrium but, at least with respect to the acid deposition issue, it is difficult to conceive of receptors and emitters forming coalitions. The equilibra I seek are therefore of the Nash type. A Nash equilibrium is a pair of values for x and y selected by independently acting receptors and emitters such that no individual has an incentive to alter his choice.

ASSIGNING THE BURDEN OF FALSE POSITIVES AND FALSE NEGATIVES

As already noted, D represents the losses society will suffer if an environmental effect that could be attributed to acid deposition is unequivocally realized. The prospect threatens emitters with increased control costs and receptors with environmental damages. These expected losses are the joint result of a policymaker acceptance of causative linkages between precursor emissions and realized impacts, as well as a policymaker decision about whether emitters and/or receptors must do something. For example, realized environmental effects from elevated acidity levels need not be blamed on acid deposition. Responsibility for the recently increased acidity of some Scandanavian aquatic ecosystems has occasionally been assigned to altered local land management practices. Similarly, even if acid deposition is declared guilty, disagreements may remain about who emitted the precursors. Assigning the blame for an event and assigning the burden of doing something about it could be treated as different decisions. For this paper, they are viewed as synonymous. The absolute and relative magnitudes of the expected emitter and receptor losses from an environmental impact are thus dependent upon the criteria policymakers employ to form cause-effect linkages and to decide who must bear the cost consequences of a formed linkage. Because the criteria influence these expected magnitudes, they will also influence, before any particular receptor or emitter activity is undertaken, the allocation decisions of these parties. Individual emitter and receptor behavior in the search for and the anticipation of gainful activities will thereby be affected. To illustrate this, consider the criteria the policymaker might employ to assign the cost consequences of a cause-effect linkage whose existence he has come to believe.

Let λ be the share ($0 \leq \lambda \leq 1$) of receptors in D conditional on the realization of these same effects. Emitters' expected total losses from a realized impact are then $D[1 - P(x,y)](1 - \lambda)$; the receptors' expected total losses are $D[1 - P(x,y^a)]\lambda$, where y^a is, given the combination of emission monitoring and penalties available for enforcement, the emission control the receptors anticipate the policymaker will enforce. See Harford [35] for a treatment of the anticipations that various combinations of monitoring and penalties would cause the policymaker to have. This anticipation determines the share of any realized damages for which receptors must be prepared to assume responsibility. The emitters' share will be determined by the emission controls they actually adopt, y. Rewriting Equation 1, the policy-maker's social-cost minimization task is then:

$$\underset{x,y}{\text{Min }} S(x,y) = cx + wy + D[1 - P(x,y^a)]\lambda + D[1 - P(x,y)](1 - \lambda) \tag{4}$$

One of the necessary conditions for the minimization of Equation 4 is

$$w = DP_y(1 - \lambda) \tag{5}$$

or the marginal cost of emission control must be set equal to the emitter share of the expected societal losses caused by a unit change in actual emission control. The level of emission control that results from meeting Equation 5 is then:

$$y^\lambda = y^\lambda(w,x,D,\lambda) \tag{6}$$

Assume that emitters are unable to surprise receptors and that the policymaker knows this. In accordance with rational expectations, the receptors perfectly antici-pate emitter maneuvers such that the systematic component of the emission control levels they anticipate is always equal to the emission control level that solves Equation 5. The latter is unfailingly obtained by the policymaker. Given this, the policymaker's interest in minimizing social cost still requires that he induce receptors to adopt cost-minimizing levels of stress resistance. With $y^\lambda(\cdot)$ substituted for y^a, the following problem is then equivalent to Equation 4:

$$\text{Min } S(x,y^\lambda) = cx + wy + D[1 - P(x,y)] \tag{7}$$

subject to Equation 5; that is, the policymaker's problem is now identical to Equa-tion 1 with the exception that he is constrained to meet the emission control level that corresponds to Equation 5. The necessary minimization conditions for Equation 7 subject to Equation 5 are:

$$c = DP_x - \frac{D\lambda P_y P_{xy}}{P_{yy}} \tag{8}$$

and Equation 5, where Equation 8 is the result of ridding the conditions for x and y of the Lagrangian multiplier. Assume $P_{xy} > 0$. This assumption says, as seems reasonable, that the incremental contribution of stress resistance to type I (false positive) error increases with increasing emission controls. The social costs, $S(x^\lambda,y^\lambda)$, resulting from fulfillment of Equations 5 and 8 will correspond to the cost-minimizing solutions, $S(x^*,y^*)$, in Equations 2 and 3 if and only if λ, the receptors' share of a realized environmental impact from acid depositions, is set equal to zero. Otherwise, since the right side of Equation 8 after DP_x is positive, the receptor will overinvest in stress reduction. Similarly, from Equation 5, the emitter will underinvest in emission controls. In effect, even when the level of emission control is optimal, socially cost-minimizing receptor behavior requires that emitters face the prospect of paying for all increased control costs and receptor losses when and if an environmental impact is realized. Full policymaker knowledge and enforcement of optimal levels of emission control will, in the absence of complete emitter responsibility for all cost consequences of realized environmental impacts, fail to ensure minimum social costs.

The results on inefficiency immediately above depend on both the emitter and receptor bearing some nonzero portion of any realized environmental impact. Similar results can be obtained where the receptors' investment in stress resistance is set at the cost-minimizing level for Equation 4, and the emitters' level of control is a response to this investment. In this case, the emitters' responses will not conform to Equation 3 unless they must bear all the cost consequences of a realized environmental event. Of course, different responses would be obtained if the constraint attached to Equation 7 failed to conform to one of the necessary minimization conditions for Equation 4. Regardless of the responses that resulted however, unless $\lambda = 0$, the combination of response and constraint must be nonoptimal since the constraint would, by reference to Equations 2 and 3, be nonoptimal.

With acid deposition, it is equally as unrealistic to presume that emitters would have to pay for any losses suffered by receptors as it is to presume that receptors would have to pay for increased emitter control costs, regardless of whether one or the other, or neither, are operating their respective activities at socially cost-minimizing levels. Receptor and emitter payments are assumed to disappear into the public treasury. Therefore, contrary to Page [24], in the absence of the perfect information that guarantees there can be no wrong policymaker decisions, an acid deposition control program which minimizes the social cost of wrong decisions is fundamentally nonattainable. Using a somewhat different route and in another context, Brown [33] arrives at a similar conclusion. Cooter [36] has recently argued that, if learning is allowed, this conclusion need not hold in a dynamic setting. However, since the source of inefficiency in the above development arises from the magnitude of λ, it is worthwhile to consider the cost consequences of variations in λ. This can be accomplished by solving Equations 5 and 8 and for x and y in terms of w, D and λ. Thus:

$$\frac{\partial x}{\partial \lambda} = -\frac{DP_y(P_y + P_{xy})}{[P_{yy} + D(P_x P_{yy} - P_y P_{xy})](1 - \lambda)} > 0 \qquad (9)$$

and

$$\frac{\partial y}{\partial \lambda} = \frac{P_y}{P_{yy}(1 - \lambda)} < 0 \qquad (10)$$

where third-and higher-order derivatives are allowed to vanish. From Equation 10, the greater the share of the expected cost consequences the receptors must bear, the fewer the efforts the emitters will make to control their precursor emissions. Receptors, as Equation 9 indicates, respond to this by increasing the defensive and adaptive measures they adopt.

Retaining the assumption that $y^a \equiv y^{\lambda}$, the overall expected social cost consequences of an increase in λ are derived by substituting Equations 9 and 10 into Equation 4, and then differentiating with respect to λ. On doing so, one obtains:

$$\frac{dS(x,y)}{d\lambda} = x^{\lambda}(c - P_x D) + y^{\lambda}(w - P_y D) \qquad (11)$$

which, from Equations 8 and 5, is:

$$\frac{dS(x,y)}{d\lambda} = x^{\lambda} - \frac{D\lambda P_y P_{xy}}{P_{yy}} + y^{\lambda}(-D\lambda P_y) > 0 \qquad (12)$$

In effect, as the receptors' share of the cost consequences increases, the social costs of acid deposition impacts and acid deposition precursor control increase despite any enhanced defensive and adaptive measures receptors may adopt.

A simple, somewhat frivolous, example conveys the essence of the above formal development. Consider a person who could save $20 in control costs by emitting some acid deposition precursors. Futher assume he could save $17 in control costs at no loss in saleable output by adopting a low-polluting technology. The polluting production process causes $15 in harm to some distant fishermen, while the low-polluting technology harms them not at all. If the emitter knows for this reason that he will have to pay for the losses he causes the fisherman, he can expect to gain no more than $5 by using the polluting technology and at least $17 by using the low-pollution technology. With payment required, he would obviously choose to adopt the latter technology since, relative to the polluting technology, he would be better off by $12. However, if he knew that no payment would be required, he would select the polluting technology since, relative to the low-pollution technology, he could expect to have $3 more in wealth. Resource allocation decisions of the victimized fishermen will be similarly distorted toward activities (e.g., watching fishing programs on television worth only 10 cents to them) rendering them less susceptible to the emitter's predations.

CRITICAL VALUES

Except insofar as they have been identified respectively with targeted emission control–stress resistance combinations and the expected cost consequences of a realized environmental effect, false positives and false negatives have played a minimal role in the preceding section. More explicit attention is devoted to them in this section, particularly with respect to the implications for emitter and receptor allocation decisions of alternative policymaker choices of critical values for testing hypotheses.

$P(x,y)$ can be looked at as an imperfect system for testing hypotheses about stochastic cause-effect linkages or about the uncertain source origins of acid deposition precursors. Allow the policymaker to employ a test statistic z $(0 < z < 1)$, which he applies to both stress resistance and emission control hypotheses and which defines $(\bar{x}, \bar{y})$ as the cutoff values representing the boundaries between the acceptance and the critical regions of the hypothesis tests. Low values of z and the cutoff values $(\bar{x}, \bar{y})$ associated with them, are employed with more discriminating or "finer" tests and, when achieved, indicate a high probability that the tested hypothesis is true. Alternatively stated, they imply that the critical region for rejecting the hypothesis is large. For any particular sampling distribution of emission control levels, for example, $\bar{y}$ is associated with the critical test value the policymaker employs for rejecting the null hypothesis of a cause-effect linkage between acid deposition and an environmental impact. Reductions in the probability of a type II error (false negative) would then require either a decline in the fineness of this test (an increase in the critical value) and/or the acquisition of more information about "true" stress resistance. Similarly, $\bar{y}$ is the consort of the critical test value the policymaker uses for rejecting the null hypothesis of a linkage at a given location between precursor emission controls at a particular set of sources and acid deposition.

Disregarding the cost of acquiring the information to test a hypothesis, consider the cost consequences for emitters (M) of alternative policymaker findings about the magnitude of measured ecosystem stress resistance, i.e., whether x is greater or less than $\bar{x}$. In effect, the policymaker must decide whether ecosystem sensitivity is great enough to display acid deposition-induced stresses. For simplicity, $\bar{x}$ is assumed to be equal to x^* the socially cost-minimizing stress resistance level earlier set forth in Equation 2. Four cases can be distinguished.

If $x < \bar{x}$, so that measured stress resistance is less than its cutoff value necessary for rejecting the hypothesis of acid deposition-indiced stresses, and if no environmental impact is ultimately realized, the emitter cost consequences are:

$$S^M(\bar{x},y) = wy + D_o[1 - P(\bar{x},y)](1 - \lambda) \qquad (13)$$

where the second term on the right-side continues to be the expected emitter cost of a prospective type II error. But, if $x < \bar{x}$, and an environmental impact is finally realized, the emitter cost consequences over the entire period are:

$$S^M(\bar{x},y) = wy + D_o[1 - P(\bar{x},y)](1 - \lambda) + D_1[1 - P(x,y)] \qquad (14)$$

In Equation 14, D_o represents the actual additional control costs and receptor loss payments emitters must now bear because an estimated link between actual precursor emissions and realized and observed environmental impacts could not be rejected. The *ex ante* D_o and *ex post* D_1 emitter cost consequences will generally differ because stress resistance is insufficient to convince the policymaker that emitters are innocent of responsibility for observed as opposed to expected receptor losses.

It might also happen that $x \geq \bar{x}$. If so, it would not matter to the emitters whether or not an environmental impact occurred because no link between their precursor emissions and observed environmental impacts would be recognized. The emitter cost consequences for both cases would be as in Equation 13.

It follows from Equations 13 and 14 that when $x < \bar{x}$, our risk neutral emitters must minimize the expected value of:

$$S^M(\bar{x},y) = wy + D_o[1 - P(\bar{x},y)](1 - \lambda) + D_1[1 - P(x,y)] \qquad (15)$$

Similarly, when $x \geq \bar{x}$, the emitters minimize the expected value of:

$$S^M(\bar{x},y) = wy + D_o[1 - P(\bar{x},y)](1 - \lambda) \qquad (16)$$

Note that in Equations 14 and 15 it is actual ($\equiv$ measured, by assumption) stress resistance x and not the cutoff stress resistance value $\bar{x}$ that determines the emitter cost consequences of a realized environmental impact.

The respective marginal conditions describing the emitters' choices of y given acceptance in Equation 17 and rejection in Equation 18 of the hypothesis of a precursor emission–environmental impact linkage are:

$$w = D_1 P_y + D_o P_y (1 - \lambda) \qquad (17)$$

and

$$w = D_o P_y (1 - \lambda) \qquad (18)$$

All terms in Equations 17 and 18 are positive, including P_y; thus since Equation 17 includes the extra term $D_1 P_y$, it is more costly than Equation 18. This extra term represents the cost consequences for the emitter of the marginal influence of an increase in emission control type I (false positive) error. Investments in emission control are costly to the emitter not only because of the direct outlays they require but also because they increase the likelihood of false positives and thereby reduce the probability of false negatives. When measured stress resistance is less than the cutoff stress resistance value required for rejecting a cause-effect linkage between acid deposition precursors and environmental impacts, emitters therefore have an incentive to invest too little in emission controls so as to reduce the probability of environmental impacts being wrongly attributed to acid deposition. The higher the cutoff value, the more frequently will the measured value of stress resistance be less than the cutoff value, and the more intense the incentive for the emitter to adopt this form of strategic behavior, given that the number of emitters is not so large that the individual emitter's control efforts have no effect on type I error. The critical region of the test is so large that a finding unfavorable to the emitter is difficult to avoid. A demonstration similar to Equations 13–17 for the receptor would, when $y > \bar{y}$, show him overinvesting in stress resistance so as to enhance the probability of false positives, thereby reducing the probability of false negatives.

Although I have sidestepped the issue by assuming that all emitters are identical, note that if some emitters can provide emission control at less cost than others, different emitters will employ different probabilities of error in their decisionmaking. Thus, as Erlich and Becker [32] point out, differences in probabilistic perspectives about contingent states may be dependent more on control costs than on personal temperament.

Clearly, the policymaker's choices of $\bar{x}$ and $\bar{y}$ will play a major role in emitter and receptor behavior. Further perusal of Equation 17 provides further insight into how these choices complicate the policymaker's decision problem. Note that if $\lambda = 1$ in Equation 17 so that the full burden of the losses from any realized environmental impact accrues to receptors, the term $(1 = \lambda)D_oP_y$ disappears. The expression then becomes identical to the social cost-minimizing condition for emitters set forth in Equation 3. Setting $\lambda = 1$, however, opposes the condition ($\lambda = 0$) for socially cost-minimizing emission controls present in Equations 5 and 8 that is also required to make Equation 18 correspond to Equations 5 and 8. Thus, the policymaker, when choosing and announcing λ, the share of the cost consequences the receptors will bear if and when an environmental impact is realized, must trade off one set of socially nonoptimal biases in emitter and receptor behavior against another set.

SUMMARY AND CONCLUSIONS

The abovementioned tradeoff in biases implies that the total social costs of acid depositions may be nonmonotonic in the combination of the receptors' share of a realized environmental impact and in the critical values policymakers and researchers use to reject or to fail to reject hypotheses about cause-effect or source-cause linkages. For example, if emitters know they will be made responsible for only a minor portion of the cost consequences of a realized environmental impact, a relatively "coarse" test for failing to reject either of the aforementioned linkages may have less total social cost than does a more discriminating test; the lesser emitter responsibility generates fewer emitter costs, while the less stringent critical values favor receptors.

All this presumes that emitters and receptors perceive their interactions with the policymaker as being contingent on less-then-complete policymaker knowledge of cause-effect and source-cause linkages. The various equilibrium conditions demonstrate that the policymaker's choices of λ, $\bar{x}$ and $\bar{y}$ directly influence emitter and receptor incentives and thus the total social costs of their activities. An interesting conclusion follows for the practice and use of scientific research in regulatory policy. In particular, in a stochastic world in which emitters do not have to bear the full cost consequences of their activities, "scientifically" less-than-ideal hypothesis test procedures can alter emitter and receptor investment incentives in desirable ways. Very lenient criteria for failing to reject the results on environmental impacts embodied in scientific reports can reduce the total social costs of acid depositions and their control. This result in no way depends on costly information acquisition for control costs or environmental impacts.

ACKNOWLEDGMENTS

John Tschirhart, Larry Regens, and the participants in a seminar at the University of Washington have straightened me out on several points. Any remaining errors are my responsibility. Partial financial support has been provided under a grant from the U.S. Environmental Protection Agency.

REFERENCES

1. Nriagu, J.O. "Deterioration Effects of Sulfur Pollution on Materials," in *Sulfur in the Environment, Vol. II*, J.O. Nriagu, Ed. New York: John Wiley & Sons, Inc., 1978, pp. 1–60.
2. Evans, L.S., C.A. Conway and K.F. Lewin. *"Yield Responses of Field-Grown Soybeans Exposed to Simulated Acid Rain,"* paper presented at the International Conference on the Ecological Impact of Acid Precipitation, Sandefjord, Norway, March 11–14, 1980.
3. Falk, D.L., and W.A. Dunson. "Effects of Season and Acute Sublethal Exposure on Survival Times of Brook Trout at Low pH," *Water Res.* 11:13–15 (1977).
4. McFee, W.W. "Effects of Acid Precipitation and Atmospheric Deposition on Soils," in *A National Program for Assessing the Problem of Atmospheric Deposition (Acid Rain),* J.N. Galloway et al., Eds. (Fort Collins, Co: Natural Resource Ecology Laboratory, Colorado State University, 1978), pp. 64–73.
5. Holden, A.V. "Sulfuric Waters," in *Ecological Effects of Acid Precipitation,* M.J. Wood, Ed. (Surrey UK: Central Electricity Research Laboratories, 1979).
6. Jefferies, D.S., C.M. Cox and P.J. Dillion. "Depression of pH in Lakes and Streams in Central Ontario During Snowmelt," *J. Fish. Res. Board Can.* 36:114–121 (1979).
7. Patrick, R.V., P. Binetti and S.G. Halterman. "Acid Lakes from Natural and Anthropogenic Causes," *Science* 211:446–448 (1981).
8. Abrahamsen, G. "Acid Precipitation Plant Nutrients, and Forest Growth," in *Ecological Impacts of Acid Precipitation: Proceedings of An International Conference,* D. Drablos and A. Tollan, Eds. (Oslo, Norway: SNSF, 1981), pp. 58–63.
9. Wiklander, L. "Leaching and Acidification of Soils," in *Ecological Effects of Acid Precipitation,* M.J. Wood, Ed. (Surrey, UK: Central Electricity Research Laboratories, 1979).
10. Brown, D.J.A., and K. Sadler. "The Chemistry and Fishing Status of Acid Lakes in Norway and Their Relationships to European Sulphur Emissions," *J. Appl. Ecol.* 18:433–441 (1981).
11. Niemann, B.L. *"Review of Major Long-Range Transport Air Quality Simulation Models,"* paper presented at the Office of Technology Assessment Mini-Assessment on LRTAP, U.S. Congress, February 1981.
12. Cogbill, C.V. "The History and Character of Acid Precipitation in North America," *Water, Soil, Air Poll.* 6:407–413 (1976).

13. Oden, S. "The Acidity Problem—An Outline of Concepts," *Water, Air Soil Poll.* 6:137–166 (1976).

14. Galloway, J.M., and G.G. Parker. "Difficulties in Measuring Wet and Dry Deposition on Forest Canopies and Soil Surfaces," in *Effects of Acid Precipitation on Terrestrial Ecosystems,* T.C. Hutchinson and M. Havas, Eds. (New York: Plenum Press, 1980), pp. 57–68.

15. Kerr, R.A. "Tracing Sources of Acid Rain Causes Big Stir," *Science* 215:881 (1982).

16. Curtis, C., Ed. *Before the Rainbow: What We Know About Acid Rain* (Alpha, NJ: Sheridan Printing Co., 1980).

17. Carter, L.J. "Uncontrolled SO_2 Emissions Bring Acid Rain," *Science* 204:1181–1182 (1979).

18. Irving, P.M., and D.A. Sowinski. "Effect of H^+, $SO_4^=$, NO_3^-, and NH_4^-, Concentrations and Ratios in Precipitation Applied to Greenhouse-Grown Soybeans," in *Annual Report, Part III,* ANL-80-115 (Argonne, IL: Radiological and Environmental Research Division, Argonne National Laboratory, 1980), pp. 6–10.

19. Abrahamsen, G., J. Hovland and S. Hagvar. "Effects of Artificial Acid Rain and Liming on Soil Organisms and the Decomposition of Organic Matter," in *Effects of Acid Precipitation on Terrestrial Ecosystems,* T.C. Hutchinson and M. Havas, Eds. (New York: Plenum Press, 1980), pp. 341–362.

20. "The Debate over Acid Precipitation: Opposing Views, Status of Research," Comptroller General of the United States, Washington, DC (1981).

21. Klopatek, J.M., W.F. Harris and R.J. Olson. "A Regional Ecological Assessment Approach to Acid Deposition: Effects on Soil Systems," in *Atmospheric Sulfur Deposition,* D.S. Shriner, et al., Eds. (Ann Arbor, MI: Ann Arbor Science Publishers, Inc., 1980), pp. 539–553.

22. Cowell, D.W., A.E. Lucas and C.D.A. Rubec. *"The Development of an Ecological Sensitivity Rating for Acid Precipitation Impact Assessment,"* Working Paper No. 10, Lands Directorate, Environment Canada, Burlington, Ontario (1981).

23. Cole, P.S. Ed. *Acid Rain: A Transjurisdictional Problem in Search of Solution,* (Buffalo: State University of New York Canadian-American Center, 1982).

24. Page, T. "A Generic View of Toxic Chemicals and Similar Risks," *Ecol. Law Quart.* 7:207–244 (1978).

25. Jacobson, J.S., "Acid Rain and Environmental Policy," *J. Air Poll. Control Assoc.* 31:1071–1073 (1981).

26. Bazelon, D.L. "The Judiciary: What Role in Health Improvement?" *Science* 211:792–793 (1981).

27. Ricci, P.F., and L.S. Molton. "Risk and Benefit in Environmental Law," *Science* 214:1096–1100 (1981).

28. Olson, M. "Environmental Indivisibilities and Information Costs: Fanaticism, Agnosticism, and Intellectual Progress," *Am. Econ. Rev. Papers Proc.* 72:262–266 (1982).

29. Neyman, J., and E.S. Pearson. "On the Problem of the Most Efficient Tests of Statistical Hypotheses," *Phil. Trans. Roy. Soc.* A 231:289–337 (1933).

30. Levi, I. *Gambling with Truth* (New York: A. Knopf & Co., 1967).

31. Winkler, R.L. *An Introduction to Bayesian Inference and Decision* (New York: Holt, Rinehart and Winston, Inc., 1972).

32. Erlich, S., and G.S. Becker. "Market Insurance, Self-Insurance, and Self-Protection," *J. Polit. Econ.* 80:623–648 (1972).

33. Brown, J.P. "Toward an Economic Theory of Liability," *J. Legal Studies* 2:323–349 (1973).

34. Crocker, T.D., and B.A. Forster. "Decision Problems in the Control of Acid Precipitation: Nonconvexities and Irreversibilities," *J. Air Poll. Control Assoc.* 31:31–37 (1981).
35. Harford, J.D. "Firm Behavior Under Imperfectly Enforceable Pollution Standards and Taxes," *J. Environ. Econ. Manage.* 5:26–43 (1978).
36. Cooter, R., J. Kornhauser and D. Lane. "Liability Rules, Limited Information, and the Role of Precedent," *Bell J. Econ.* 10:366–373 (1979).

CHAPTER 6

Normative Economics and the Acid Rain Problem

L.S. Eubanks
R.A. Cabe

Acid rain is only one of several environmental threats of great concern today that seem to pose serious challenges to economic analysis. A list of the characteristics of the acid rain problem that are particularly troublesome would include: (1) transboundary effects; (2) extreme uncertainties; (3) potentially irreversible impacts; (4) intertemporal impacts; (5) episodic impacts. Although this list is certainly substantial, the purpose of this chapter is not to analyze any of these specific characteristics. Rather, this chapter intends to provide the perspective necessary to evaluate the relevance of normative economics to the consideration of policy alternatives available for responding to environmental problems such as acid rain.

Looking at normative economics in perspective seems especially timely. More frequently these days policymakers at all levels are asking economists for direction in choices affecting the pattern of resource use in the economy. Certainly the interest in economics is welcome. However, it is feared that in the process of interaction between economists and public policymakers, qualifications that are essential to placing economic analysis in its proper perspective may go unstated, be overlooked, or be misunderstood. If this occurs, the consequences are very likely to be substantial and undesirable. The reason for this is related to the fact that normative economics has defined its scope in such a way that the policymaker is expected to play a substantial evaluative role, yet policymakers seem today to rely ever increasingly on economic evaluations to set the direction of public policy. It is hoped that this chapter can help provide the perspective on normative economics necessary to judge its merits for public policymaking.

SCOPE OF NORMATIVE ECONOMICS

Normative economics is the area of economics concerned with making policy-relevant statements. Normative economics stands in contrast to positive economics, which is concerned with description, explanation and prediction. The concern in normative economics is with "optimal" use of a society's resources. This concern involves not only identifying and describing optimal resource uses, but also evaluating the available means for achieving an optimal use of the available resources.

Normative economics has attempted to limit its scope in a special way. If we consider the classic distinction between ends and means, normative economics has limited its scope to the study of the means to attain a given end. According to this tradition in normative economics, economists as economists are unqualified to make arguments for or against the acceptance of specific normative criteria for judging socially optimal resource use. The emphasis has been on allowing philosophers and politicians the domain of choosing the normative criteria that will define what is meant by "optimal" in the economist's study of the optimal allocation of a society's endowment of resources. As such, the role of the normative economist is to provide a special expertise which is concerned with identifying the means by which any given concept of optimality in resource use can be obtained. Mishan [1] provides a simple definition of normative economics that expresses this tradition:

> Normative or welfare economics can be defined as the study of criteria for ranking alternative economic situations on the scale of better or worse.

Actual practice of normative economics seems not to have lived up to this tradition, nor fulfilled this definition. Virtually the entire body of normative economic analysis has been concerned with only one criterion for ranking alternative economic situations, this criterion being the concept of efficiency due to Pareto. Furthermore, the definition of efficient resource use, in theory and in application, is quite specific in nature, and suggests a specific normative framework or ethic for evaluating desirable economic outcomes and actions.

Underlying all of normative economics, to such an extent as to suggest a definition of the economic method, is the individualistic perspective: that the basis of value is the set of choices or subjective valuations of the individual members of society, and nothing more. Individual preferences are taken to be exogenous indicators of relative values for society. Value to society only arises from individual valuations.

Given this basic maxim of normative economics, the criterion for judging optimal (efficient) resource use is the criterion of Pareto optimality: "A situation is Pareto optimal if it is impossible to make one person better off except by making someone else worse off" [2]. This normative criterion is used in theoretical analyses to identify conditions which are necessary for an efficient pattern of resource use. It can clearly be seen as an efficiency criterion by considering a situation which would be inconsistent with it. Surely, if it would be possible to change the pattern of resource use in production and/or consumption in a way which would make at least one person in society better off without at the same time harming another person, then this initial situation must be inefficient.

Normative economists realize that there are a number of weaknesses associated with using the criterion of Pareto optimality to identify desirable patterns of resource use. However, acknowledging such weaknesses has, in general, not led the normative economist to search for alternative criteria, since the acknowledged weaknesses are consistent with the defined scope of normative economics itself. The two major weaknesses acknowledged by normative economists are: (1) the Pareto criterion only provides a partial ranking of resource use patterns, and (2) the Pareto criterion can identify patterns of resource use as optimal which may be regarded as unjust on distributional grounds. In other words, rather than identifying a single efficient (desirable) allocation of resources, the pareto criterion generally identifies an infinity of allocations as efficient and therefore optimal, some of which are sure to be regarded as unjust. Choices among pareto optimal allocations, however, are left for politicians and philosophers to make, presumably on distributional grounds or with the use of additional criteria for optimality.

The criterion of Pareto optimality is useful in theoretical models of normative economics, but not very useful in empirical applications. In actual situations it would be hard to imagine making any change from the status quo that would improve one person's position without provoking the complaints of at least one other person. The operational version of Pareto optimality is the criterion of potential Pareto improvement. When considering a change from the status quo to a new situation, the new situation is regarded as a potential Pareto improvement (and desirable) if those who benefit from the change can more than compensate those who are harmed by the change. Clearly, if these conditions hold, then making the change is a Pareto improvement and consistent with the criterion of Pareto optimality, given those made worse off are actually compensated.

Normative economists also realize weaknesses in the use of the potential Pareto criterion in evaluating policy alternatives. Perhaps the major problem area is that the operational criterion for judging optimal resource use only requires the potential for compensating the losers of any change and not actual compensation. In effect, the operational version of normative economics does not require that an actual Pareto improvement be attained in order to offer policy evaluations and policy prescriptions. This position is justified, again, by reference to the traditional scope of normative economics. Economists as economists are not qualified to decide whether compensation must be made. Such choices are left to philosophers and politicians.

One possible rationale for the fact that economists have restricted their attention almost exclusively to the Pareto criterion (and its operational variant, the potential Pareto criterion) is the belief that any criteria that are likely to be proposed by philosophers or politicans will be consistent with the Pareto criterion—will choose among the infinite number of Pareto optimal allocations but will not deny a change desired by at least one person and harms none. In this rationale the Pareto criterion is regarded as a minimal property that will apply to a broad class of criteria that might be suggested by philosophers and politicians. An economist who applies the Pareto criterion offers minimal policy prescriptions that should be consistent with any "reasonable" view of social welfare, and the economist has not violated the province of philosophers and statesmen by advocating one "ethic" over another.

This "safe" strategy for the research program of normative economics is unacceptable because it is based on a false premise. The Pareto criterion is *not* consistent with many conceptions of social welfare and this divergence becomes especially relevant when normative economics is extended to the evaluation of policies with dramatic environmental consequences and very long time profiles such as those that arise in consideration of the control of acid precipitation.

The inadequacy of restricting attention to the Pareto criterion is even more fundamental. The individualistic perspective that underlies the Pareto criterion rules out consideration of a variety of factors that seem to play a large role in actual political deliberations on policy alternatives. Whether a policy is likely to be conducive to social stability, whether it might foster the preservation of widely held values, whether it will tend to limit or expand individual freedom, access to the political process, social mobility, etc., are all arguments which can carry weight in actual social decision-making but which can have no intrinsic value under an individualistic perspective of social welfare. These considerations can enter into a conventional cost/benefit analysis but their roles can only be instrumental— acting through individuals' utilities. When these considerations arise in the political process, however, we do not find statesmen arguing that a reduction of personal liberty entailed by one policy alternative will reduce individual utilities. Instead, it will be argued that reduction of personal liberty is an intrinsically bad thing, to be avoided in public policy. This implies that an individualistic perspective is not appropriate for the evaluation of some types of public policies.

A nonindividualistic argument that has a long history in U.S. political rhetoric is that of granting intrinsic value to the welfare of a generalized "posterity." The individualistic perspective denies validity to a general posterity but *may* allow explicit consideration of future individuals. Here the individualistic perspective admits an ambiguity. If some consequences of a public policy will fall on members of a future generation, the question arises whether or not these future people are to be accorded the full status of "individuals" under the individualistic perspective of normative economics. Marglin [3] judged that they should not: "I want the government's social welfare function to reflect only the preferences of present individuals." This is a narrow version of the individualistic perspective that would have us reject arguments other than individuals' utilities and, furthermore, explicitly consider as individuals only those living today. This resolution of the ambiguity of the individualistic perspective with respect to future people does not exclude the possibility that the prospective welfare of future people enters into the utilities of present individuals, thus giving instrumental value to the welfare of future generations. However, it does deny intrinsic value to the welfare of future people.

The way in which distant future effects are handled is of vital importance to the evaluation of most environmental changes and in particular to the evaluation of public policies with regard to acid precipitation. Two admirable efforts in this area by normative economists deserve special mention because they do live up to the definition of normative economics by examining alternative criteria for ordering social states. Page [4] examines the implications of the "present value criterion" and the "conservation criterion" and Page [5] examines the "Kantian ethic" in the context of normative economics. Schulze et al. [6] consider the implications of the "utilitarian ethic," "libertarian ethic" and "democratic ethic" for evaluation of the question of proper storage of nuclear wastes. These papers are mentioned in passing as being exemplary of the type of research effort that follows when normative economists devote their attention to the examination of alternative criteria proposed by philosophers and statesmen.

HELP FROM PHILOSOPHERS

If the scope of economics relies on philosophers and politicians for the normative criteria that define what "optimal" resource use is, then it should be interesting to examine the views of philosophers on questions of interest to normative economics. In what follows the analyses of several philosophers will be discussed. The discussion focuses on the implications of specific philosophical analysis for normative economics in general, but also with respect to the positions of philosophers and normative economists regarding problems such as acid rain in particular.

Recently, there has been a growing interest in ethical philosophy with environmental ethics. Obviously, environmental ethics should be relevant not only to the evaluation of public policy on the environment, but also to an evaluation of economic activities and the patterns of resource use related to the environment. Perhaps the normative framework used by economists in evaluating environmental policy does/does not command wide acceptance among philosophers.

Regan [7] suggests a categorization that is interesting and that may be useful in clarifying the nature of the ethic implied by the normative economic framework. Regan discusses the possibility of constructing an environmental ethic by attempting to describe conditions that are necessary and sufficient for such an ethic to exist. In doing so, Regan suggests that attempts to support a particular version of an environmental ethic can be divided into management theories and kinship theories.

Management theories propose an ethic for the use of the environment that suggests that the environment ought to be used in a way that enhances the quality of human life, including possibly the quality of life of future generations of humans. The way in which Regan gives attention to future generations is interesting. It suggests not only that it is important to be concerned in our contemporary choices with the position of future generations, but also that the precise nature of the concern is as yet undetermined within philosophy. This should be comforting to some resource economists who were perhaps concerned with questions of intergenerational justice even before the ethical philosophers. Perhaps more important, however, is the feeling of some ethical philosophers "that the question of our obligations to the future can be seen as a litmus test for an ethical theory" [8]. We might wonder if there is a similar statement to be made with regard to normative economics.

Kinship theories expand the set of interests given moral consideration to nonhuman animals because they are kin to humans in the fundamentally importance way of being conscious. The distinction between management theories and kinship theories with respect to nonhuman animals is exactly analogous to the question discussed above of whether future humans are to be granted intrinsic or merely instrumental value.

This categorization squarely identifies normative economics as a management theory which presents an ethic for the use of the environment. Management theories direct us to preserve wildlife, for example, only if this is in the interest of human beings, possibly including consideration of the interests of future humans. Normative economics identifies a specific definition of what is to be regarded as the interests of human beings, i.e., the willingness to pay to obtain something of value. As such, normative economics suggests we ought to take an action or actions to preserve wildlife threatened by acid rain only if doing so will provide greater benefits than costs to the human species, possibly with the inclusion of the benefits and costs to future generations.

In contrast, kinship theories require that wildlife interests be regarded on their own right. Policy evaluation on the basis of a kinship theory would require an explicit additional computation of the interests held by wildlife in actions to limit acid precipitation, perhaps including the interests of future wildlife. Unfortunately it seems that the kinship ethic does not make clear how to carry out the required computations for the interests of wildlife itself.

Regan regards a kinship ethic as a necessary condition for an environmental ethic (or "ethic *of* the environment"). A genuine environmental ethic, in Regan's view, holds "that the class of these beings which have moral standing includes but is larger than the class of conscious beings—this is, all conscious beings and some nonconscious beings must be held to have moral standing" [7]. Therefore, any genuine environmental ethic would consider the interests of human beings, wildlife and some nonconscious beings such as trees. Presuming the existence of such an environmental ethic, determining whether an obligation exists to limit acid rain (partially or completely) would require consideration of the intrinsic interests (and perhaps future interests) of humans, wildlife, and trees.

These distinctions between management ethics, kinship ethics and environmental ethics are useful in placing the propositions and policy prescriptions of normative economics in perspective. They allow the economist as economist to properly qualify her policy suggestions with respect to the problems posed by acid precipitation as well as other environmental problems. Accepting, for example, that philosophy can derive a kinship ethic or even an environmental ethic has the implication that policy suggestions based on normative economics will be incomplete, and perhaps entirely without an acceptable normative basis, because both of these ethics depart from the individualistic perspective which is fundamental to normative economics as presently constituted.

However, it is important to note that this categorization of ethics suggests a serious difficultly. Suppose an action designed to limit acid rain implies a conflict between the interests of humans and the interests of other conscious and nonconscious beings, where the net value of the human interests is negative while that of the other nonhuman beings having moral standing in the issue is positive. How are we to adjudicate the conflicting interests? How are we to weigh the calculation of interests for each type of being having moral standing? This problem is not dissimilar to problems discussed in economics involving interpersonal utility comparisons or of comparing alternative Pareto optimal states, and unfortunately the philosophers have had as much difficulty as economists finding acceptable solutions. Although arguments suggesting kinship ethics and environmental ethics may be accepted, the bases for such arguments have not been capable of also suggesting a solution to the problem of weighing conflicting claims of the different types of beings considered to have moral standing.

Economists should not be overly quick to use this problem as a point of attack asserting that the economic ethic is the only point of view that allows adjudication of conflicting interests (even if only human interests are regarded), and therefore is the only point of view allowing the possibility of making determinate evaluations of alternative actions. Pointing out the difficulty is not to say the difficulty is unresolvable. More importantly, it may be the case that compelling reasons can be given for rejecting the view that only human beings have moral standing. Regan [7] actually argues that compelling reasons can in fact be given for rejecting arguments that limit moral standing only to conscious beings.

It is also possible to distinguish between two types of ethical theory. These categories are discussed in Sayre [9]. One type of ethical theory is the theory of moral obligation. This theory is concerned with responsibilities we hold as members of society. The second type of ethical theory is concerned with what is good and with the nature of such goods.

The first type of ethical theory should also be categorized into teleological and deontological approaches. According to teleological theories the moral status of an action is to be evaluated by the good produced by the action. Examples of teleological theories are utilitarianism and egoism. Utilitarianism holds that the obligation of moral agents is to produce the greatest good for the greatest number, while egoism contends that all moral agents are obligated to maximize their own individual long-term wellbeing. In contrast, deontological approaches regard the moral status of an action as being independent of specific beneficial consequences. Major deontological theories are contractarianism and Kantianism. According to contractarian theories a moral agent's obligation is to act according to principles of justice that would secure the fairest distribution of goods within the society. Kantianism holds that a moral agent should act according to maxims that one is willing to make universally binding.

Note that the second type of ethical theory, while being concerned with what is good, is often also concerned with the nature of goods that are to be maximized in teleological theories. Specifically the good to be maximized in teleological theories has often been regarded as satisfaction, pleasure or economic utility.

On the basis of these distinctions, the position of normative economics can clearly be considered teleological rather than deontological. Furthermore, normative economics seems at times to be aligned with both the utilitarian and egoistic positions, accepting the principle that it is desirable to use the available resources to produce the greatest good for the greatest number (utilitarian), and also the principle that the greatest good is discovered by reference to each individual's satisfaction or utility.

This individualistic perspective of normative economics which takes the only basis of value as being individual preferences also places normative economics opposite the second category of ethical theory, which is concerned with identifying what is good. Rather than accept value as being revealed by exogenous individual utility functions, this second ethical framework would find it perfectly reasonable to evaluate the preferences themselves as good or bad. The obvious contrast to be drawn is that perhaps such an alternative ethical framework would regard some actions which are harmful to the environment, e.g., acidification of lakes that results in the killing of fish populations, as unacceptable while the ethic of normative economics regards the action as acceptable.

It is not the intention of this chapter to support one position over another. Such intentions would be contrary to the traditional scope of normative economics. Rather, the chapter hopes to provide some perspective on the most frequently analyzed framework for normative economics, and to suggest that normative economists should be true to the definition of normative economics and begin to examine alternative criteria for evaluating resource uses.

By way of example, the widely read theory of justice provided by Rawls [10] will now be examined. Rawls uses the approach taken by contract theory, which assumes that society is a cooperative venture undertaken for mutual advantage. However, conflict of interest still occurs in such a framework because individuals are not indifferent to outcomes or to the set of advantages to which they have claim. Thus, a set of principles is necessary that can choose between alternative societal institutions that define the way in which the advantages and burdens of social cooperation will be determined and distributed.

For Rawls, an institution is a public system of rules that defines offices and positions and their associated rights and duties, benefits and burdens. By a "public system of rules" it is meant that every person taking a part in the cooperative venture of society knows these rules and knows that all the others in the society know these rules which define the basic institutions in the society. As such, institutions influence an individual's life prospects, the expectations an individual has about what she might be, or might be able to do in her life. In this way institutions are seen to favor certain starting positions over others. Since no individual can be said to deserve the place in society to which she is born, nor is necessarily deserving of the natural abilities she acquires at birth, the point of social justice is to develop a set of institutions that would not favor such inequalities in starting positions but would instead lessen the extent by which the starting position for an individual is important in determining her life prospects.

Rawls' framework is based largely on the notion of pure procedural justice. The idea is to design the structure of the society so that the outcome is considered to be just regardless of whatever the specific outcome happens to be. The key to understanding this concept is to realize that an outcome can be considered just even when there is no criterion for independently judging the right result. As an example, consider the outcome of a World Series baseball game between the Dodgers and the Yankees. The rules by which the outcome of the game is to be determined are well defined. These rules are also accepted by both parties to the baseball contest, and both parties are committed to follow the rules and abide by the outcome that results when the rules of the game are impartially carried out. In such a circumstance, the outcome is just, regardless who is declared the winner, as long as the rules of the game are equally and impartially administered. It would be inappropriate to apply a separate criterion to the outcome of the game and determine that the Dodgers rather than the Yankess should be the winners. Rawls wants a similar framework in order to evaluate the justice of a society's distribution of benefits and burdens of social cooperation. If the set of rules or basic institutions that distribute the benefits and burdens of social cooperation are judged to be just, then the specific outcome which results will be just regardless of the nature of the outcome.

Turning to the Rawlsian principles of justice themselves, in discovering the principles of justice Rawls appeals to a theoretical construction referred to as the "original position." Rawls considers the original position as a sort of assembly of representatives for all parties to the social contract which is set apart from any actual society. (The theoretical nature of the concept "original position" must be emphasized. The concept in no way is meant to suggest an actual gathering of representatives. The purpose of the concept is to allow abstract discussion of a set of principles of justice.) The subject of discussion by this hypothetical assembly is an agreement on a conception of social justice, i.e., a set of principles that will guide the development and arrangement of society's major institutions. To ensure that the assembly in the original position is a fair meeting between morally equal persons the nature and extent of information and knowledge available to the assembly's members is restricted. The idea is that each person in the original position is to be deprived of all morally relevant information, i.e., all information that would enable an individual to attempt to tailor the principles of justice to her own advantage. In such a position each individual would appear to be willing to accept principles of justice that would be mutually advantageous to all.

What principles would individuals so situated choose to guide the way in which the major social institutions are chosen and arranged? Rawls of course argues that his principles of justice (which are summarized in Table I) are the principles that would be chosen. Note that these principles are meant for choosing between alternative social institutions, not alternative allocations of resources to specific individuals.

The principle 1 is self-explanatory. The Rawlsian principles of justice are lexically ordered with the principle 1 having first priority. As such, no set of basic economic institutions would be consistent with the principles of justice if all individuals did not have equal liberty.

Turning to the principle 2, Table I indicates that principle 2b, the principle of fair equality of opportunity, is to have priority over the difference principle,

Table I.   The Rawlsian System of Justice

Principal 1: each person is to have an equal right to the most extensive total system of equal basic liberties compatible with a similar system of liberty for all.

Principle 2: social and economic inequalities are to be arranged so that they are both: (a) to the greatest benefit of the least advantaged, consistent with a just savings principle, and (b) attached to offices and positions open to all under conditions of fair equality of opportunity.

Priority Rule 1 (The Priority of Liberty): the principles of justice are to be ranked in lexical order and therefore liberty can be restricted only for the sake of liberty. There are two cases: (a) a less extensive liberty must strengthen the total system of liberty shared by all; (b) a less than equal liberty must be acceptable to those with the lesser liberty.

Priority Rule 2 (The Priority of Justice Over Efficiency and Welfare): the second principle of justice is lexically prior to the principle of efficiency and to that of maximizing the sum of advantages; and fair opportunity is prior to the difference principle. (2a) There are two cases: (a) an inequality of opportunity must enhance the opportunities of those with the lesser opportunity; (b) an excessive rate of saving must on balance mitigate the burden of those bearing this hardship.

General Conception of Justice: all social primary goods—liberty and opportunity, income and wealth, and the bases of self-respect—are to be distributed equally unless an unequal distribution of any or all of these goods is to the advantage of the least favored.

which is principle 2a. Note that principle 2b focuses attention to inequalities as they are attached to "offices and positions" rather than to particular individuals. This point relates primarily to the emphasis that Rawls places on institutions. Institutions define a set of rules for the distribution of benefits and burdens as they are attached to social offices and positions rather than to specific individuals. The relevance of this point to normative economics is that the preferences of specific individuals may not be of relevance. Rather, concern should perhaps be directed to an index or a set of descriptive characteristics that identify inequalities as they are attached to social positions. If it is desirable to retain the use of a utility index, then it may be necessary to attach the utility index to a representative individual holding a specific position.

Looking at the difference principle, it is important to note the phrase: "Social and economic inequalities are to be arranged. . . ." The idea is to begin from the reference position of equality between the advantages attached to each social position. Certainly it would be difficult not to accept an equal distribution of benefits and burdens. Yet, social cooperation is thought to have advantages for all. Therefore, if the situation is such that inequalities are beneficial to all, this also should be acceptable from the point of view of justice since all members of the society are benefited by the particular arrangements of social cooperation, even if the benefits and burdens are distributed unequally.

In effect, Rawls accepts a commonly held view that individuals may have special abilities and talents that will be to the benefit of all if cultivated in the society. To motivate the use of special talents and abilities, it is acceptable to offer a greater number of advantages to positions that require a greater development of special talents and abilities. The presumption is that such talents properly developed and applied can be to the benefit of all individuals in the society.

In what way are the Rawlsian principles of justice relevant to the questions and models of normative economics? This question will be examined in two contexts. In the first context, the relevant question will be concerned with evaluating the standard normative economic framework generally, while the second context will be concerned with whether the Rawlsian framework can be used to say anything about environmental problems such as acid rain in particular.

One interest of normative economists has been the construction and analysis of mathematical representations of an economy. Using such models to identify optimal resource uses requires the statement of the Pareto criterion in a mathematical problem of maximizing one person's utility function subject to the constancy of the levels of utility of all other persons. It would seem interesting to try to construct a similar type of analysis using the Rawlsian framework.

Of all the principles of justice listed in Table I, the principle most amenable to such an exercise probably is the difference principle. At least on the face of it, the mathematical statement of the Rawlsian system in an economic model would be simple and straightforward. Assuming that the principles of equal liberty and fair equality of opportunity are satisfied, a mathematical statement of the Rawlsian principles is simply to maximize minimum utility associated with persons in the society. An example in which such a mathematical statement can be found is Solow [11]. However, such a representation is probably oversimplified and naive.

First, in the context of the typical normative economic model, such a mathematical model will tend to imply the conclusion that resources should be used in a fashion that equalizes utilities or welfare levels for all persons in the society. Surely, such a conclusion is in general not consistent with the nature of Rawl's theory of justice. The difference principle is itself intended to identify justifiable social and economic inequalities. A mathematical formulation that implies equality is too much simplified to be able to express the richness of the Rawlsian framework of justice. A more complex formulation is necessary, and might be found by specifying constraints to a max-min problem much as is done with the Pareto criterion. Such constraints might involve ranking the persons in society by utility levels (of course implicitly necessary to at least a limited extent to identify the least well off in the society), or perhaps some sort of "transformation function" that expresses the way benefits and burdens are allocated to the least well off position as inequality increases. It would appear that a simple application of the Rawlsian system to normative economic models will not be accomplished in terms of a simple restatement of a mathematical objective function. A more complete discussion of the problems involved in relating the Rawlsian difference principle to static normative economic models can be found in Eubanks [12].

Second, normative economic models are essentially concerned with evaluating specific allocations of resources to specific persons. It has already been noted that the Rawlsian system is quite different in this respect, and is meant to be applied to the basic institutions in the society. The Rawlsian system, following the notion of pure procedural justice, is concerned with the choice of the fundamental "rules of the game" that determine the way in which the benefits and burdens fall on the members of society. The specific allocations to specific persons are not the subject of justice nor the subject of the Rawlsian principles.

Noting this point suggests a problem for normative economics. The vast majority of normative economics has been constructed to examine specific allocations, and the vast majority of normative economists accept the traditional scope of normative economics in which the framework for evaluating resource use is the domain of philosophers and politicians. Here is a philosophical framework that suggests that the subject of normative economics should not necessarily be specific allocations to specific individuals. Furthermore, the fact that explicit analysis of alternative institutional situations has not been a frequent topic of normative economics indicates that conventional normative economic analysis has little relevance for social decision making under this framework.

The problems for normative economics as commonly practiced are compounded when it is realized that the philosophical framework of Rawls provides a lexical ordering of the importance of each of the principles found in Table I. In particular, the principle of equal liberty has first priority, then the principle of fair equality of opportunity, the difference principle, and the principle of efficiency. The principles of justice have priority over efficiency. Social institutions are to be first arranged in a just fashion before it is appropriate to consider questions of efficiency.

Normative economics has been concerned almost entirely with questions of efficiency, limiting itself to the extent that it regards questions of justice to be outside its domain. Normative economics, viewed from the perspective of Rawlsian justice, has defined its domain to be only a limited subset of all the considerations relevant to evaluating the optimality or desirability of a pattern of resource use. This discussion is not intended to discard the work or normative economics. It is intended to put this work in perspective given the domain of analysis accepted by normative economics itself, and to suggest that there is a wealth of work to be done in the "study of criteria for ranking alternative economic situations on the scale of better or worse." Schotter [13] is a promising work in these respects. Schotter is concerned with an analytical framework for the explicit analysis of institutions and actually favors the following as a definition of normative or welfare economics: "In essence, welfare economics is the study of the optimal rules of the game for economic and social situations." (p. 6)

Is the Rawlsian system of justice relevant to analyzing the problem of acid rain and policy responses designed to deal with acid rain? Suppose we regard the economy as generating both the benefits and burdens of social cooperation. The benefits of social cooperation can be thought of as the goods and services produced with the society's resource endowment, while one of the types of burdens can be thought of as pollution. Given the materials balance principle formulated by Kneese et al. [14], *all* activities of production and consumption are seen as generating residuals that must ultimately be returned to the environment. Social cooperation, through the economic activities of production and consumption, generates burdens in the form of residuals that must be dealt with by persons in the society. Therefore one of the basic burdens of social cooperation is the cost associated with residuals and pollution. Furthermore, institutions such as markets and common property will determine where the burdens of pollution will fall in the society. A complete discussion of this framework can be found in Eubanks and Meuller [15].

If this view is accepted, then the Rawlsian framework suggests that a relevant question is whether the burdens of pollution problems such as acid rain are distributed in a just fashion, i.e., in a manner consistent with the Rawlsian principles of justice? Another question of relevance would be whether the institutions associated with alternative public policies with respect to acid rain are consistent with the Rawlsian principles of justice?

Explicit answers to such questions will not be provided in this chapter. However, it is possible to suggest some considerations and potential research directions that seem interesting. The first consideration, and perhaps the most important, is to note that, even assuming the principles of equal liberty and fair opportunity are satisfied, economic analysis of the Rawlsian principles will be required to identify the least well off position in the society. In addition, it must be possible to relate the impact of alternative policy alternatives and the implied institutional frameworks in the society to improvements or decrements in the least well off position.

Certainly this is no simple task. However, economics has long concerned itself with positive questions concerning income distribution, both by describing distributions empirically and by explaining the forces at work in determining income distribution. Actually from a normative Rawlsian point of view, the entire income distribution under alternative institutional arrangements is of little relevance. The concern is with increments and decrements in the position of the least well off in the society.

Given a way of identifying the position of the least well off in the society, then perhaps the next step would be a variant of cost/benefit analysis. Only in this case the relevant concern is again limited to only the least well off in the society. Whether the magnitude of the empirical task would be lesser under a Rawlsian analysis than a standard cost/benefit analysis is not clear. Standard cost/benefit analysis evaluates increments and decrements of the welfare of all members of the society, not just the least well off. But such evaluations can very often be accomplished by reference to markets for which information is easily available. The Rawlsian analysis would require attention to be paid specifically to one group and not to all groups in some general aggregate.

Some of the relevant analysis might also be available in the economics literature on incidence, specifically the incidence of pollution remedies such as charges and standards, and on the property value approach to valuing nonmarket commodities. See Freeman [16] for a review of the literature on this subject. Consider the hypothesis involved in the property value approach to nonmarket valuation. Values associated with polluted air and clean air will be capitalized into property values. As such it seems reasonable to conclude that the institutional arrangements defined by common property and the market system work out to impose the greatest burden of social cooperation in the form of air pollution on the least advantaged in the economy. Such an inference does not guarantee that the existing system of institutions violates the Rawlsian principles of justice, since the least advantaged also share in the benefits of the actions that generate the air pollution. The question concerning the net advantage or disadvantage to the least well off must be left for a more detailed analytical and empirical analysis. All of the analytical detail may not be a part of the bag of tricks of normative economics today, but it certainly seems that providing the remaining analytical details will be a tractable research problem.

Since this section has briefly considered the implications of Rawlsian justice for cost/benefit analysis, it seems appropriate to conclude with one final viewpoint found in the philosophical literature with implications for cost/benefit analysis. The normative basis for cost/benefit analysis is the potential pareto improvement. It was pointed out above that economists recognize that this criterion may be a weak basis for policy statements, since it only requires potential compensation, but that the question of whether compensation should be paid is for philosophers and politicians. What are the positions of philosophers?

Nozick [17] has stated that "if a polluting activity is to be allowed to continue on the ground that its benefits outweight its costs (including its pollution costs), then those who benefit actually should compensate those upon whom the pollution costs are initially thrown." On Nozick's view then the potential Pareto criterion is an insufficient basis for applied normative economics with regard to pollution. Cost/benefit analysis provides an answer to the question, "does the change at hand represent a potential Pareto improvement"? This answer is arrived at by determining the total amount of compensation required by those who would be harmed by the change (costs) and comparing this amount to the sum that those favoring the change would contribute to assure that the change occurs (benefits). If the benefits exceed the costs then compensation is hypothetically possible and the potential pareto criterion has been met.

In order for the change to be desirable in Nozick's view it must represent an actual Pareto improvement and a great deal more is required than meeting the test of a conventional cost/benefit analysis. Under Nozick's framework the cost/benefit analyst must make a detailed analysis not only of aggregate benefits and costs but must determine to whom compensation is due and from whom this compensation can be extracted. Furthermore, a scheme of collections and disbursements must be devised that will be acceptable to all parties concerned so as to make the change an actual Pareto improvement. The costs of devising and administering this scheme of payments and compensations become costs of the project and must be reckoned along with the other costs in evaluating the desirability of the change.

The important point is that by requiring only hypothetical rather than actual compensation we not only ignore the distribution of costs and benefits but we change the structure of costs. The word "hypothetical" in the potential Pareto criterion allows us to ignore the (potentially large) class of transaction costs that must be evaluated if we seek an actual Pareto improvement.

CONCLUSION

This chapter has defined the scope of normative economics in the now traditional way as the study of criteria for evaluating alternative economic situations. The scope of normative economics is limited in a way that excludes arguments that favor or oppose specific criteria for judging optimality. Politicians and philosophers are regarded to have expertise to make such arguments.

It is unfortunate that standard practice of normative economics is almost exclusively associated with the study of only one criterion, the efficiency criterion. The implication of this is that normative economics as practiced has a very limited normative basis.

This conclusion is illustrated clearly in this chapter when a number of contributions to the philosophy literature are examined. It is not suggested that normative economics is of little use in policymaking, only that the self-delimited scope of normative economics should be understood in its proper perspective. The practice of normative economics has not been entirely true to its traditional definition because it is seldom interested in the study of criteria posing as alternatives to the criterion of efficiency. Certainly there is an enormous amount of interesting normative economic analysis, as traditionally defined, concerning alternative normative criteria. Normative economic analysis certainly has relevance to policymaking with respect to environmental problems such as acid rain, but perhaps only if it remains true to its defined scope and allocates its effort to the analysis of a much larger set of criteria for evaluating desirable resource uses.

REFERENCES

1. Mishan, E.J. *Introduction to Normative Economics* (New York: Oxford University Press, 1981).
2. Layard, P.R.G., and A.A. Walters. *Microeconomic Theory* (New York: McGraw-Hill Book Company, Inc., 1978).
3. Marglin, S.A. "The Social Rate of Discount and the Optimal Rate of Investment," *Quart. J. Econ.* 77:95–111 (1963).
4. Page, T. *Conservation and Economic Efficiency* (Baltimore, MD: John Hopkins University Press, 1977).
5. Page, T. "A Kantian Perspective on the Social Rate of Discount," Social Science Working Paper 278, California Institution of Technology, Pasadena, CA (1979).
6. Schulze, W.D., D.S. Brookshire and T. Sandler. "Economics and Ethics?: Evaluating the Risks of Storing Nuclear Waste," *Natural Resources J.* 21:321–340 (1981).
7. Regan, T. "The Nature and Possibility of an Environmental Ethic," *Environ. Ethics* 3:19–34 (1981).
8. Bickham, S. "Future Generations and Contemporary Ethical Theory," *J. Value Inquiry* 15:169–177 (1981).
9. Sayre, K. "Morality, Energy, and the Environment, *Environ. Ethics* 3:5–18 (1981).
10. Rawls, J. *A Theory of Justice* (Cambridge, MA: Harvard University Press, 1971).

11. Solow, R.M. "Intergenerational Equity and Exhaustive Resources," *Rev. Econ. Studies,* special supplement issue on The Economics of Natural Resources (1974).
12. Eubanks, L.S. "Normative Economics and the Rawlsian Theory of Justice," Working Paper Series, Economics Division, University of Oklahoma (1982).
13. Schotter, A. *The Economic Theory of Social Institutions* (Cambridge: Cambridge University Press, 1981).
14. Kneese, A.V., R.U. Ayers and R.C. d'Arge. *Economics and the Environment: A Materials Balance Approach* (Baltimore, MD: Johns Hopkins Univ. Press, 1970).
15. Eubanks, L.S., M.J. Mueller. "Pollution, Economics and Rawlsian Justice," Working Paper Series, Economics Division, University of Oklahoma (1982).
16. Freeman, A.M. *The Benefits of Environmental Improvement: Theory and Practice* (Baltimore, MD: Johns Hopkins Univ. Press, 1979).
17. Nozick, R. *Anarchy, State, and Utopia* (New York: Basic Books, 1974).

CHAPTER 7

Economic Impact of Acid Precipitation: A Canadian Perspective

Bruce A. Forster

The purpose of this chapter is to provide a Canadian view of the economic implications of the acid precipitation phenomenon in North America. Noneconomists are frequently surprised to learn that economists are interested in studying the phenomenon generally called acid rain. Presumably, this is because noneconomists feel that this is in the domain of chemists (who know about acids), meteorologists (who know about rain) and natural scientists (who know about plants and things that need water). Why, in this natural order, they wonder, would an economist intrude? The tongue-in-cheek answer is that economists as a professional group are sufficiently imperialistic to feel that their special skills enable them to say something about virtually any subject (provided that the human race is in any way interested in the subject).

The serious answer to why economists are interested in studying the acid rain phenomenon is very much a consequence of the general principle enunciated by the tongue-in-cheek answer. Economics is the study of the allocation of the resources of society and the implications of the various allocations for the welfare of the members of society. The acid rain phenomenon that is currently concerning society is generated within the economic system as the result of various production and consumption activities. Power plants and smelting refineries produce commodities that society desires. The by-products of these production activities are the precursors of acid rain. Consumption of automobile services will produce precursors of acid rain in addition to travel services. Changes in the level of these activities and/or changes in the nature of these economic activities will in general have implications for the amount of acid rain that is associated with anthropogenic sources.

If society wishes to cut back on the emission of acid precursors it will either face reduced output levels of the commodities that are associated with the emissions or it will have to incur higher costs of production, which will imply reductions of outputs elsewhere in the economy.

If acid deposition occurs, it has various potential impacts on the economy. These effects are important to economists because the members of society value the items that may be affected. The natural resources such as fisheries and forests that may be affected yield commercial products as well as recreational and esthetic products desired by the members of society. Society is therefore confronted with a trade-off between various types of commodities.

The above paragraphs are designed to explain why economists as social scientists are interested in the acid rain phenomenon. The interest of the economist, however, goes beyond descriptive analysis. The economist is interested in acid rain as an issue for economic policy. The acid rain phenomenon is an example of what economists call an externality, spillover or external effect of economic activity. Externalities are basically unintentional side-effects of a given activity. When these unintentional side-effects result in a reduction in someone's welfare, we speak of an external diseconomy or a negative externality. Economic theory tells us that when an activity generates a negative externality, a free market may allow too much of that activity relative to the socially ideal level.

The acid rain phenomenon is an example of a very pervasive negative externality. It can be discussed in terms of individual economic agents since the activities of producers and consumers affect the wellbeing of other producers and consumers. The phenomenon can be discussed as a case of regional externalities. As a result of the long-range transport of air pollutants, activities in one region of the country may affect another region some distance away. A natural extension of the regional externality framework is to consider the international or transboundary externalities. Economic activities in both Canada and the United States give rise to external effects across international boundaries. The impact of current rates of deposition may be felt well into the future. Thus, the acid rain phenomenon may be considered as an intergenerational externality. Many of the impacts are irreversible, so that current activity may permanently reduce the natural resource wealth that we pass on to future generations.

As mentioned above, a free market may lead to excessive levels of activities that generate acid precursors, since those responsible for the activities will not take proper account of the costs they impose on others. This is true when those "others" are individual consumers and producers, regions, sovereign nations or future generations.

The official Canadian point of view concerning acid rain is that the phenomenon is the joint concern of Canada and the United States. Steps in the direction of cooperation have been seen in the establishment of the United States–Canada Consultation Group on the Long-Range Transport of Air Pollutants and the signing of the Memorandum of Intent between the governments of the United States and Canada, concerning transboundary air pollution.

The first section of this chapter will review theoretical frameworks for discussing the economic issues discussed in the foregoing paragraphs. There are two key features that the development needs to take into account: (1) the international nature of the acid rain phenomenon; and (2) the dynamic nature of the phenomenon that causes the welfare of future generations to be affected by current actions.

Thus formulated, the policymaker is confronted with an optimization problem that needs to be solved across nations and through time. To solve this problem, the theorist raises various ethical considerations that must be confronted. These issues will be considered in the first section of the chapter.

The second section will review the available information on the impacts of acid deposition on selected Canadian resource industries to determine their potential economic significance. To determine the economic impact, the economist first needs to know the dose-response relationship, which measures the responsiveness of resource outputs to varying levels of acidity. Once the economist knows the response, he is in a position to attempt to place monetary values on the foregone outputs. In some cases, market data may be used (subject to various qualifications). In other cases, no market data are available and the economist must resort to alternative techniques to reveal economic values.

ECONOMIC THEORY FOR ANALYZING TRANSBOUNDARY POLLUTION

The purpose of this section is to review some existing economic models that deal with environmental and natural resources issues in an international or intertemporal framework. These models will be assessed in terms of their contribution to the transboundary pollution problem.

Asako [1] discusses a model of environmental pollution and international trade in commodities. His interest is in the interaction between pollution and the pattern of trade as it affects social welfare. Social welfare depends positively on the consumption level of two commodities and negatively on the level of environmental pollution generated by the production of the two commodities. Production and consumption are constrained to satisfy the balance-of-payments constraint that exports must equal imports in value terms.

Asako shows that, for an economy with decentralized decision-making, expanding trade may reduce the country's welfare if the export sector is pollution-intensive. If the import-competing sector is pollution intensive, then expanding trade always increases welfare since the usual gains from trade will result and domestic pollution will decrease. Asako concludes "that a (small) country can and should transfer indirectly its pollution problem to foreign countries by controlling international trade activities." He points out that a superior result would involve a tax-cum-subsidy policy to restructure the domestic economy in the pre-trade position.

The analysis by Asako ignores the pollution level in the foreign country. Pethig [2] considers a similar problem but explicitly introduces the foreign country into the analysis. Pethig demonstrates that the country that specializes in the pollution-intensive commodity for export will suffer a welfare loss while its trading partner will always gain from trade. This result assumes no active pollution emissions control.

When both countries adopt environmental controls, the pattern of trade may be influenced by the relative restrictiveness of the control policies. The country with the less restrictive policy will export the pollution-intensive commodity. Neither country loses from trade provided the emissions control constraints were binding in the pretrade position.

As Pethig points out the level of each country's welfare depends on the emission standards in the other country so that the issue of interdependent welfare maximization is raised. While the degree of control in one country affects the other, the actual level of emissions in a given country has an impact on that country only. Welfare is interdependent in Pethig's framework because the emissions standards affect the pretrade structure of the economy. In Asako and Pethig, welfare levels are interdependent since the reorganization of production as a result of trade has implications for the pollution levels in each country. A country can, in effect, "export" pollution by importing pollution-intensive goods, although the pollution itself is not physically transported between countries.

While these models are useful for analyzing the restructuring of international production in light of environmental pollution, they do not provide an accurate representation of the long-range transport of pollutants issue, since the total amount of pollution in each country is determined entirely by its own production levels. Markusen [3,4] allows the level of pollution in a given country to be a function of the level of production of one commodity in each of the two countries engaged in trade.

Pollution is then treated as a pure international public bad. This formulation goes some way to being useful for the transboundary problem, since some pollution is beyond the control of the domestic authorities. The Markusen framework can be used to shed light on the recent debate concerning the export of power by Ontario Hydro to the United States. Ontario may benefit even though the added production could cause acid precipitation in Ontario. If the increased Ontario production merely replaces foreign production of acid precursors, it is possible that the overall amount of acid precipitation could remain constant. Ontario then reaps the usual "gains from trade" from being able to export the power.

In Markusen's framework the behavior of the foreign country is not considered explicitly. Its presence is shown through the trade and pollution effects. However, optimizing behavior is left to a national government, which has the responsibility of maximizing a social welfare function. Braden and Bromley [5] explicitly recognize the behavior of the foreign country in a model that is polar to that of Pethig and Asako. The two countries share a common level of pollution but do not engage in the trade of commodities. Each country produces a single consumption good that generates a flow of pollution that enters a common environmental media. Social welfare in each country depends on its own consumption level and the aggregate amount of pollution determined as the sum of pollution from the two countries as in Markusen's model.

This model is useful for considering the transboundary pollution issue. Each country would be better off if the other country reduced the level of its polluting activities. The global optimum is characterized by the condition that the sum of national marginal rates of substitution between consumption and pollution must equal the marginal rate of joint production association. This last expression is defined as the rate of increase in pollution given an increase in consumption and is similar to the marginal rate of transformation for goods and services. The authors are interested in comparing the global optimum with the outcomes that might result from various forms of negotiation.

The authors consider a framework in which each country could determine an optimal strategy treating the foreign level of emissions as predetermined. The national optimal consumption level then becomes a function of the foreign emission level. Treating this as a varying parameter permits the analyst to trace the country's reaction path. Each country will have such a reaction path. These two reaction paths intersect at the Cournot equilibrium, which "is the best that independently optimizing parties can hope to achieve unless explicit efforts are made to cooperate." The indifference curves that pass through the Cournot equilibrium define the maximal utility levels that can be achieved by independent optimization. The area bounded by these curves contains all points which are Pareto superior to the Cournot point. The points that satisfy the global optimum will lie on the contract curve in the interior of this area. These points can be achieved only with cooperation between the two sovereign nations.

The authors examine possible tax schemes that would allow the two countries to reach the contract curve. However, explicit use of a global social welfare function is not made; hence, the question of the distribution of net gains from cooperation is left open. Markusen [3,4] and Braden and Bromley [5] suggest that it may be necessary to consider transfer payments in order to produce a Pareto-optimal outcome.

The Braden and Bromley model is useful for transboundary pollution issues because of the focus on the environmental issue between two sovereign nations. It does not consider the trade aspect that Asako, Pethig and Markusen consider in their treatments; however, these authors do not treat the sovereign nations on an equal basis dealing with a common pollution problem. Each approach offers some insight to various aspects of the problem.

None of the approaches discussed thus far considers the possibility that the pollution effects could be cumulative over time. In particular, lake systems or soils may become acidified over periods of time. In this case the level of pollution (acidification) is dependent on past actions. The models discussed above are static models in which the pollution is a flow dependent only on current activities.

To deal with this aspect it is necessary to formulate a dynamic model in which the stock of pollution can vary over time. Asako [1] discusses such a model in the context of international trade. A flow of waste generated by current production activity is added to the stock of pollution, which is subject to a decay process that for acidification could be representative of buffering action or flushing of the waterway. The objective of the central planners is to maximize the discounted value of future utility.

As in his static model Asako shows that trade should be curtailed (promoted) if the pollution-intensive commodity is the export (import) good. Labor should be shifted from the pollution-intensive good to the less intensive sector over time if the pollution level is below its long-run desired value. The reverse is true if the initial pollution level is higher than the long-run value. These policies are self-centered and may not be in the spirit of "good neighborliness."

In Asako's dynamic framework the pollution is generated only by domestic activities. Pollution is exported by reorganizing production internationally. Forster [6] considers a model that is a dynamic analog of the Braden and Bromley model. In this framework the stock of common pollution is generated by the individual consumption levels in the two sovereign nations.

Unlike the Braden and Bromley static version the Forster framework allows for active antipollution activities. This permits consideration of pollution control devices to reduce emissions of acid precursors or liming of lakes to reduce acidity. The decay function is specified so as to allow for a complete loss of buffering ability in a waterway.

This framework considered a specific method of dealing with the question of joint international optimization. The method requires that one country maximize its own welfare subject to ensuring that the other did not fall below some specified minimum level. The optimal terminal state is analogous to that obtained by Braden and Bromley. The result is adjusted to account for the existence of discount rates on future welfare and decay of pollution. It is shown that due to the special form of the decay function, multiple equilibria may exist. Many other dynamic models such as Asako's produce single equilibrium results.

The model does not provide a mechanism for specifying the minimum level of welfare for one of the countries. Presumably, neither party would enter into a collective agreement if it believed it would achieve a level of welfare lower than it would achieve by refusing to cooperate. By treating the minimum level as a parameter it is possible to generate a utility possibility frontier.

The models discussed thus far have all treated pollution as an esthetic problem insofar as it affects utility directly but does not have an impact on the production capacity of the economies concerned. Hence, indirect utility effects operating through impacts on commodities or natural resources are not considered explicitly. These models may be appropriate for effects such as visibility reductions that have been noted in eastern North America.

Forster [7] discussed a descriptive model of growth and pollution for a single economy not engaged in trade for which the level of pollution reduced the productive capacity of the economy. In this model it was possible that the system could spiral into an equilibrium, continually cycle about the equilibrium, or result in an ultimate collapse of the economic system (assuming no corrective actions were ever taken).

Cropper [8] considers a case in which there is a threshold pollution level at which a catastrophe occurs, reducing utility to the level that would occur if consumption was zero. In a sense this allows for a feedback that is not merely esthetic in nature. However, it only occurs if the threshold is hit. This model may be appropriate for some aspects of the transboundary pollutants where major problems do not occur for some levels of pollution but severe problems could occur for some uncertain levels of pollution. Cropper shows that the optimal solution may produce multiple equilibria.

Crocker and Forster [9] construct a model to describe the generation, transport, deposition and resource impacts of long-range transported pollutants such as acid precursors. It is assumed that the waste is generated in one region and travels to another region where a portion is deposited. The waste deposited in L accumulates as a stock and this accumulated stock of waste has an impact on resources R in region L. The environmental carrying capacity of the resource $\bar{R}$ is thus a function of the level of pollution $\bar{R} = \bar{R}(P)$. This may be interpreted as a dose-response relationship that relates pollution levels to natural resource levels. The authors discuss the importance of the nature and shape of this function for economic analysis. In particular they discuss problems of nonconvexity and irreversibility of the dose-response relationship. These features are important for applied work as will be discussed below.

The Crocker and Forster model is intended as descriptive framework. The model allows for long-range transport that could cross political boundaries, but these boundaries are not explicit. To solve an international optimization control problem at this level of generality would be quite difficult. If this model is interpreted as one for transboundary pollution, then it is implied that the flow of pollution is unidirectional from one country to another. To reflect more accurately the North American situation the flows would have to occur in both directions. It would be necessary to consider local emission sources as well as local resource impacts. There is work currently underway to establish transfer matrices which will associate emitting regions and receptor regions [10].

To discuss the optimal control of acid deposition or other pollution problems it is necessary to specify a criterion by which alternative policies may be ranked. First, consider a single country attempting to determine an optimal plan over time. Many authors assume that the objective is to maximize the present value of the discounted flow of future utility as done by Asako [1]. This is one objective function but it may not be the only one nor may it be the correct one.

Some would argue that the choice of a social objective function is an ethical choice. One position [11] holds that to employ a positive discount rate to discount the welfare of future generations is "ethically indefensible." Maximizing an integral of future utility (discounted or otherwise) implies that social welfare across generations is additive. Thus, redistributions of welfare from the future generations to the present are justified provided the present generation benefits more than the future generations lose. This viewpoint may not be acceptable to everyone—particularly future generations.

There has been interest in recent years in applying a concept of social justice that has been established by Rawls [12]. Rawls suggests that individuals (in principle) be asked to step behind a "veil of ignorance" in deciding on resource allocations. Since an individual is ignorant of the allocation he would receive when the process is completed it is suggested that this approach will lead to an allocation of resources that will maximize the welfare of the least advantaged (worst off) individual (group, region country or generation). This is called the maximin criterion.

This criterion seems to suggest that welfare should be constant across generations. Hartwick [13] has shown that in a natural resource framework if resources are depleted the capital stock must be augmented to compensate for this reduction. A result such as this seems relevant to the possible degradation of the natural resource base from acidification. Asako [14] discusses the issue of economic growth and pollution using the max-min principle and shows that the dynamic behavior of this system is similar to the descriptive framework of Forster [7] with a stable equilibrium, a limit cycle or an unstable equilibrium being possible. In general, when the max-min principle is feasible Asako finds that it works to retard capital accumulation and reduce pollution, thereby maintaining a higher quality of the environment.

Asako [14] considers a single nation making plans over time and hence the concern is with intergenerational equity. The Rawlsian principle may also be relevant for specifying the nature of the welfare optimization between sovereign nations. In this case it is necessary to maximize the welfare of the worst-off nation. This tends to equalize some measure of welfare between the nations. It specifies the distribution of gains in a fashion not done by the previous papers that have been discussed.

ECONOMIC EFFECTS OF TRANSBOUNDARY POLLUTANTS IN EASTERN CANADA

To determine the appropriate level of pollutants, it is necessary to compare the benefits of reducing the amount of pollutants with the cost of reducing pollutants. The costs of control are generally expressed in monetary units and are usually the dollar costs of installing appropriate devices to reduce emissions. The benefits of controlling pollutants are generally thought to be the damages avoided by having a lower level of pollution. These may include a variety of physical impacts on agricultural crops, forests, aquatic resources, health and visibility and manmade structures. The economist is usually asked to put these various physical effects on a common basis. This usually means trying to assign a dollar value to the physical changes. This dollar value for benefits can then be compared with the dollar value for control.

As discussed above, the problem of acid precipitation and transboundary air pollutants in general is an international one. Measures of the benefits of controlling these pollutants must include effects on both sides of the border. The purpose of the remainder of this chapter is to outline the information concerning the potential economic significance of acid precipitation for eastern Canada. Figure 1 [15] shows that a large portion of eastern Canada receives precipitation with pH 5.0. Southern Ontario and parts of Quebec have precipitation with pH < 4.5 (which is more than 10 times as acidic as normal clean rain).

To determine the economic value of the damages caused by pollutants, the economist could use information concerning the dose-response relationship that

Figure 1. Precipitation acidity (pH) distribution in Canada [15].

relates varying levels of pollutants to physical damage. These physical damages must then be converted into economic measures. The common measuring rod is the dollar value. In principle this is straightforward. Information on the dose-response function is supplied by the physical and biological science studies. These physical measures may then be converted into economic measures using techniques from economic analysis, which have varying degrees of sophistication.

The most appropriate procedure would use the dose-response relationship to determine a supply response for a particular commodity. This could be used to determine a price response caused by the initial physical response. The economist then determines values for changes in consumer and producer surpluses. The net change will measure the change in economic welfare. This approach would require a detailed study of specific commodity groups and is beyond the scope of this chapter, which seeks to be more broad-based. In many areas the dose-response information is not yet sufficiently conclusive to permit estimation of monetary values with confidence at this point. Crocker and Forster [9] have outlined the significance of the nature of the dose-response relationship for economic analysis.

In the following subsections the current knowledge concerning dose-response relationships is summarized and the economic significance of various receptor categories is assessed for eastern Canada.

Terrestrial Economic Effects

Only experimental evidence exists for direct effects of acidic precipitation on vegetation. Field studies have not been conclusive [16]. It is possible to observe harmful effects, but it is also possible to observe beneficial effects because the precipitation contains plant nutrients [16]. There also appears to be a lack of correlation between yield reductions and visible injury [17].

The possible response to acidic precipitation depends on a variety of factors, such as the concentrations of acids, plant species and cultivars, and the pattern and timing of the rain applications, indicating that each species may have its own unique response to the beneficial and detrimental effects of acidic precipitation [18]. Experimental evidence suggests that the sulfate and nitrate concentrations and their ratio in the acid solution may be important in plant response to acidity. Experiments have found increases in the yield of radishes when the acid solution had both sulfuric acid and nitric acid in a 2:1 ratio, while decreases in the yield of radishes when only sulfuric acid is used [17].

In some cases, detrimental effects are observed but for "unrealistically low" pH. Harcourt and Farrar [19] found that the growth of radishes appears as good at pH of 3.5–4.5 as it does at 5.5. Depressed weight was observed at pH of 2.5. In general, it appears that in experimental studies the test solution must have a pH of ≤ 3.5 to show injury [17].

Over the period 1957–1977, nine commonly grown field crops in Ontario have shown an average increase in yield of 1.88 t-ha^{-1} [20]. The specific results are shown in Table I. Two of the nine, white beans and oats, show small decreases

Table I. Changes in Yield (t-ha^{-1}) of Field Crops in Ontario: 1957–1977 [20]

Crop	1957	1977	Change
Grain Corn	3.62	5.90	+2.28
Silage Corn	22.5	29.0	+6.50
Soybeans	1.70	2.60	+0.90
White Beans	1.16	0.67	−0.49
Winter Wheat	2.22	3.50	+1.28
Barley	2.11	2.60	+0.49
Oats	1.86	1.80	−0.06
Mixed Grain	2.24	2.40	+0.16
Hay	4.45	6.10	+1.65
Weighted mean			+1.88

in yield over this period, while the others show increases. These yield changes may be the result of a variety of factors. However, it is often suggested that during this time period the acidity of precipitation has increased greatly. The results in Table I do not demonstrate that acidity has not had an impact. They do demonstrate that any negative impacts of increased acidity have been offset by other positive effects for seven of the crops. Nitrogen fertilizer recommendations for Southern Ontario have increased from 76 kg-ha^{-1} in 1964 to 97 kg-ha^{-1} in 1977 [20]. It is possible that the nitrogen compounds in acidic precipitation could have contributed to yield increases. A detailed statistical analysis would be needed to separate out the individual contributions of various factors to yield responses. These factors could include crop management techniques, fertilizer use, weather conditions and genetic changes in crop varieties. While in general the yields have increased, these increases may be lower than they might have been in the absence of increased acidity of precipitation.

However, it seems unlikely that acid precipitation is adversely affecting agricultural crops in Ontario. Ormrod et al. [21] survey the effects of various air pollutants on agricultural crops in southern Ontario and suggest that the most widespread injury to crops is caused by photochemical oxidants or oxidant smog, of which ozone is a main constituent. Various crop species are sensitive to ozone and experimental evidence suggests that the effects of acid precipitation and ozone in combination may be more than the additive effects of each individually for some species [18]. The two crops in Ontario that showed decline in yields between 1957 and 1977, white beans and oats, are known to be sensitive to ozone. The 1980 estimated farm value of all crops that are sensitive to ozone was almost $1.5 billion [22]. Increased levels of acidity in precipitation could increase the risk of damage to these crops if the combined effect of acidity and ozone is more than additive.

The above discussion points out that the scientific knowledge concerning the dose-response relationship is not conclusive enough to permit the economist to proceed with confidence. There is an alternative approach that does not use the dose-response relationship explicitly. In this approach the economics researcher studies the agricultural sector directly. The farm is the basic laboratory and the experiment is based on the environmental economics of the firm. The economist can study the cost structure of the operating farms to determine the variability in costs that are attributable to acid deposition.

To illustrate the principle in simple terms consider the following hypothetical example. Suppose we have two farms, each in a different geographic location. Suppose that the farms are identical and that all environmental conditions in the two regions are identical except for the acidity of the rainfall. In this case if there is any observed difference in the operating costs of the two farms for any given level of output then the difference may be attributed to the effects of the different levels of acidity. If acidity reduces crop yield, then the cost per unit of crop output will be higher in the region receiving the higher level of acidity. If crop yields are increased by acidity, then cost per unit will be lower for the high-acidity-receiving farm.

In the real world, farms will not be identical nor will the environmental factors affecting the production. In general, the costs of the farm will be affected by all of these factors. The multiple regression techniques of econometric analysis are designed to separate out the various contributions to cost. In principle, such econometric work should be able to assign a coefficient to acidity that would represent the incremental cost effect of acidity. This then becomes a monetary measure of the impact of acidity.

The natural science studies suggest that in this economic approach it may not be sufficient just to include data on pH of precipitation. It appears that it is important to consider the sulfate, nitrate and ozone measures also. Further, the mean pH level may not be as relevant as the peak pH during certain phases of the growing season. The economist in designing the econometric analysis needs to know what environmental factors to focus on as explanatory variables. However, in this approach he does not require the parameter estimates of the dose-response function-only the relevant variables.

This approach should work in principle. It has the attractive feature that it uses actual agricultural statistics and avoids the problem of requiring reliable estimates of parameters in the dose-response function. The knowledge gained in controlled experiments is important in defining relevant variables to be included but the economic research could begin before "hard estimates" have been determined. The two approaches could be conducted together with each serving as check or validation of the other perhaps.

There may be practical problems with this particular approach. If the relevant factors do not exhibit sufficient variation across the farm units (a possibility with acid rain) then the multiple regression techniques will be unable to detect the impact on cost. This does not mean that there is no impact, but merely that the regression techniques have assigned the impact to the constant term along with other nonvarying influences. Since acid rain may occur over a large geographic region with only small variations in pH, this may be a problem.

A simple general example of this issue occurs when acid precipitation affects the soil conditions and hence has an indirect effect on agricultural crops. It is felt that the effects of fertilization and liming will exceed the effects of acid precipitation on soils so that the effects would be minimal even where the greatest impacts are expected [16]. The fact that increased soil acidity can be offset by the application of lime does not mean that the agricultural costs of acid precipitation can be determined by examining only the direct effects on crops. The cost of the necessary liming is a cost of acidity in addition to any direct effects.

A comparison study for the Parry Sound District of Ontario for the years 1960 and 1978 revealed that the pH data were about the same [17]. Podzolized sandy soils with pH <5 in the upper horizons showed no significant decrease over the period. Linzon et al. [17] suggest that this may imply "that acidified soils are not readily altered by acidic precipitation."

Low yield spots in Ontario have been found to have lower soil pH levels than surrounding areas. It appears that crop growth is affected at a soil pH of 4.7 according to Ketcheson [20]. He further suggests "that there should be little problem with deficiencies in lime or nutrients since aside from cost, the situation can be readily corrected by applied materials." Based on soil tests at the University of Guelph, the recommended limestone treatment in Essex, Kent and Lambton counties has risen from 0.50 t-ha^{-1} in 1964 to 0.58 t-ha^{-1} in 1977. In eastern Ontario the recommended treatment has increased from 0.39 t-ha^{-1} over this period. For all southern Ontario, however, the recommended treatment has fallen from 0.48 to 0.44 t-ha^{-1} [20].

The average farm use of lime [23] is 2 t-ac^{-1}. This is likely to be the average for farms that actually use lime, since soil conditions in southern Ontario are such that many areas do not need any lime at present; however, continued acidification could result in an increased acreage requiring liming to offset the effects of acidification [24]. In soil tests at the University of Guelph for 1981, 10% of the samples required lime [24].

The Canadian cost of agricultural lime in 1980 was roughly $17 per tonne exclusive of the costs of application [23]. According to Ketcheson [20], there are 3,331,000 ha in 15 principal field crops in Ontario. This is close to 10 million acres. Suppose that only 10% of this is subject to increased acidity and in need of an additional 0.2 t-ac^{-1} of lime to counteract the increased acidity. This would lead to 0.2 × 10^6 t being required. At $17 per tonne, this is $3.4 million of lime that would be required.

In Nova Scotia, agricultural and soil scientists do not appear to be concerned about the impacts of acidic precipitation. Underwood [25] suggests that the additional lime requirements due to acid precipitation would be slightly less than 10% of the current levels. It is estimated that the additional lime costs would be $125,000.

According to MacLean [26], over half of Quebec soils have a pH <5.5. He estimates that soil erosion is removing 1.2–1.5 × 10^6 t of limestone annually from agricultural soils and a further 0.3 × 10^6 t are removed by acid precipitation. At $17 per tonne this implies that the cost of offsetting the impact of acid precipitation would amount to $5.1 million. MacLean points out that only 0.4 × 10^6 t are currently being replaced in total, suggesting that the soil resource is subject to mismanagement.

As with agricultural crop productivity, it appears that acid deposition may cause either increases or decreases in forest productivity. The result appears to depend on the composition of the acid dose and the nutrient status of the region. If the nutrient cations are abundant but the supplies of sulfur and nitrogen are low, then acid precipitation could be beneficial. If the situation is reversed, then acid precipitation may be harmful to forest productivity due to the leaching of cations from the soil.

It is possible that net beneficial effects of acid rain could occur in the short run but give way to net harmful effects in the long run due to the leaching of nutrient cations offsetting the beneficial effects of additional sulfur and nitrogen. Since trees are perennial plants there is also more concern regarding the long-run effects of repeated exposures to acid precipitation and ozone than in the case of agricultural crops.

Nitrogen deficiencies are common in the boreal and temperate forest regions.

Sulfur deficiencies appear to be less than those of nitrogen. A total 82% of the Canadian forested area is in the boreal region [27]. Responses to nitrogen fertilizer have been noted in studies concerning Canadian forests [18].

A Swedish study found evidence of a decline of 0.3–0.6%-y^{-1} in forest growth, which would imply a reduction of 30–60% over a rotation. However, these results have not been replicated [28]. Studies have associated better growth of jack pine in Minnesota, Wisconsin and northern Ontario with soils that might be classified as sensitive to acid precipitation [18].

Despite the uncertainties associated with dose-response relationships between acidity of precipitation and forest productivity, Morrison and Sullivan [28] offer the following appraisal:

> It may be stated that the general restriction of commercial forest production to "Less productive" sites, in combination with new harvesting technology (full tree harvesting) serving to reduce the availability of nutrients for recycling, and the tradition of not applying lime, may increase the vulnerability of long term forest growth to acid precipitation.

This issue is very important for Canada. About 37% of Canada's land area is forested. Approximately 42% of the acid precipitation impingement area is boreal forest, of which a third is commercial forest [15]. The forest based industries have gross sales of $16 billion per year for the nation as a whole. In the eastern provinces within the impingement area, the value is $4–5 billion annually [28]. For Quebec, the most highly sensitive and valuable forest vegetation is situated in the most heavily impacted zones [29]. Effects on forests are also of concern to Ontario, British Columbia, New Brunswick and Nova Scotia. However, airborne sea salt spray may be a significant neutralizing factor, offsetting the precipitation acidity in the Maritime Provinces [15].

Aquatic Economic Effects

There appears to be more conclusive knowledge concerning the effects of acid deposition on aquatic ecosystems. It is suspected that the effects can be traced to the sulfate ion as the driving force behind the hydrogen ion export with nitrate playing a rather minor role in Ontario [30].

There appears to be various effects of acidity on fish species. These effects include mortality, reproductive failure, reduced growth, skeletal deformity and increased uptake of heavy metals [31]. The susceptibility of fish to acidity appears to be species-specific [18].

Acute fish kills have been corrleated with waters having low pH. This may be related to elevated levels of aluminum caused by the acidification process. Levels of aluminum that are toxic to fish in laboratory experiments have been found in the Muskoka–Haliburton region of Ontario [18].

Loss of fish life from lakes may be attributed to reproductive failure since adult fish are more resistant to low pH [31]. The threshold pH levels for reproductive failure is species specific. Table II [32] presents the thresholds for reproductive failure in the La Cloche Mountain lakes of northern Ontario determined by Beamish [32]. Lake trout have become extinct in 27 lakes in the Sudbury–Temagami area due to acidification [18].

The growth of fish may decrease or increase with acidification. The growth increase is attributed to less competition for food supplies. Skeletal deformities have been noted only for white suckers [31].

The increased uptake of heavy metals could have serious implications for the health status of humans eating fish. It has been observed that the mercury content in fish is higher in acidified lakes [31]. Over a four-year period the average mercury content of Lake Muskoka trout increased from 0.5 to 3.0 mg-kg^{-1} [31]. Bass in six of nine lakes studied in 1980 had average mercury concentrations that exceed the Canadian guideline for unlimited human consumption [18].

The effects of acidification on fish stocks have important implications for commercial fisheries, sport fishing and tourism and recreation in general.

Quebec has a large number of acid sensitive lakes but the data are not available to assess the extent at this time. However, it has been observed that angling success in one area north of Quebec City declined by 30% between 1970 and 1978 [18]. Low pH has caused several rivers to be unsuitable for Atlantic salmon in Nova Scotia. Those that are no longer suitable and those that are threatened account for 30% of Nova Scotia's salmon potential, which is 2% of the Canadian potential [33]. Angling success in Nova Scotia has declined for a sample of rivers with pH <5.0 while there has been no significant change for those in the sample with pH >5.0 [33].

In Ontario, the area for which results are best known is the LaCloche Mountain range near Sudbury, Ontario. The damage was initially believed to be attributable to local sources of emissions from the smelters in Sudbury [34,35]. It has been determined that there are at least 140 lakes in Ontario that are fishless with a total of 48,000 lakes that are considered to be sensitive to acidification [36]. These numbers are constantly being revised.

The total value of the Ontario commercial fish harvest in 1979 was about $26 million [39]. However, the bulk of this harvest is from the Great Lakes, which

Table II.   Approximate pH at Which Fish in the LaCloche Mountain Lakes Stopped Reproduction [32]

pH	Species
6.0–5.5	Smallmouth bass, walleye, burbot
5.5–5.2	Lake trout, troutperch
5.2–4.7	Brown bullhead, white sucker, rock bass
4.7–4.5	Lake herring, yellow perch, lake chub

are believed to be well buffered and hence not sensitive to acidification. The catch in the northern inland waters was $2.2 million [37].

For Ontario, the sport-fishing industry is more significant in economic terms than is commercial fishing. The 1979 estimated value of expenditures for this industry was $620 million [38]. The sport fishing industry is an important factor in the tourism and recreation industry in Ontario. Of all Ontario day-use recreation occasions in 1980, 27% involved fishing activities [39]. Annual expenditures by all travelers in Ontario amount to over $6 billion in 1980 [39].

In 1977–1978 annual fishing licences were purchased by 465,212 nonresident non-Canadian anglers. An annual licence cost $10.75. A further 164,755 nonresident non-Canadians paid $6.00 each for a three-day licence; 5,612 nonresident non-Canadians paid $2.00 for organized camp fishing licences; and 24,961 Canadian, nonresident of Ontario anglers paid $4.00 each. Residents of Ontario and nonresidents who are under 17 years of age but accompanied by a licensed family member do not require a licence to fish [40]. Foreigners were willing to pay over $6 million in 1977–1978 for the privilege of fishing in Ontario waters. Non-Ontario Canadians were willing to pay roughly $100,000 to fish in Ontario.

According to Liddle [41], the fishing lodge industry generates $150 million in northern Ontario and employs 60% of that region's labor force. He estimates that annual losses of $28 million in direct and indirect economic payments could result if those fisheries were lost.

A major study is being conducted for the Ontario Ministry of the Environment to assess the economic significance of acid precipitation on Tourism and Outdoor Recreation in Ontario [42]. The results of this study are not yet available.

Haines [31] has suggested that chemical neutralization through liming may be an economically feasible remedy if the only adverse effect of acid rain is the acidification of surface waters. According to a study cited by Haines the liming costs in New York state ranged $55–470-ha^{-1} with the average being $150-ha^{-1}. The annual cost of liming in New York was estimated at $5 million. According to Bengtsson et al. [43], the Swedish costs of spreading one ton of lime was $50–70. The cost is slightly higher if a helicopter is used. Currently, they estimate 40–75 kg-ha^{-1} is required annually to compensate for acidification.

For Canada, *Harrowsmith* [23] suggests that the cost of liming would range $4000–40,000 per lake per application. According to the results of liming studies of the Ontario Ministry of the Environment, the Ontario cost figures range $50–500 ha^{-1} with 0.5 t-ha^{-1} of lime being required [44]. For lakes in region 1 of Figure 2, the $50-ha^{-1} figure is probably reasonable for most of the area because of road access in this region. The surface area of susceptible lakes in region 1 is roughly 484,000 ha [36]. Using $50-ha^{-1} for the whole area yields an estimate of $24.2 million. An application is good for three years of buffering. This cost estimate does not include the annual costs of monitoring the lakes. This could amount to $2 million per year. Thus in region 1 the annual cost of liming and monitoring would be roughly $10 million.

If the same low cost figure is assumed for regions 2 and 3, then the figure more than triples. Given the more limited access to these regions, this would

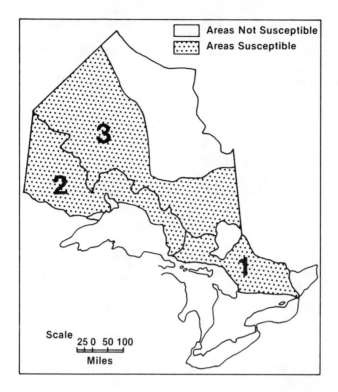

Code	Total # of Lakes	Mean Area/Lk. (Acres)	Percentage Susceptible	# of Lks. Susceptible	Total Lk. Area Susceptible (sq. mi.)
1	40,589	58.9	50%	20,295	1,868
2	76,728	98.0	20%	15,346	2,350
3	64,133	130.0	20%	12,827	2,605
Totals:	181,450			48,468	6,823

Source: Ontario Ministry of the Environment [36]

Figure 2. Susceptibility of Ontario lakes to acid precipitation effects [36].

very much underestimate the actual costs. Some "back of the envelope" figures suggest that to lime all susceptible lakes in Ontario and Quebec could produce a cost of $300 million for three years of buffering or $100 million on an annual basis [44]. This cost must be borne in perpetuity if liming is chosen as the remedial action.

If liming is chosen it may not work as planned. Metal concentrations may stay at levels that are toxic to fish gills. According to Bengtsson et al. [43] aluminum is "highly aggressive to fish gills in the pH range 4.5–6 and liming has even killed salmon and trout when the aim was to save the fish." According to Harold Harvey [45]:

You can't bring an acid lake back to the way it was by liming it. By liming an acid lake, all you end up with is a lake that was once acidic with lime in it but with a very different water chemistry and a very peculiar biota as a result.

Things may be better for lakes that are not yet acidified. Even in these cases it may be necessary to consider restocking of some fish species. This cost would be added to the cost of liming.

Intangible Economic Effects

The previous sections of this chapter have considered aspects of acid deposition that affect commodities that pass through the marketplace to some extent. Some aspects of the acid deposition phenomenon affect items in forest ecosystems, aquatic ecosystems and the rest of the economic system that do not normally pass through the marketplace. These items, however, may have value to society and a consideration of the effects on these items is required in a complete study of the benefits of controlling pollution.

Some researchers are skeptical concerning the determination of these intangible values. An extreme example of this skepticism occurs in the discussion of cultural values by Work Group 1 of the Memorandum of Intent [18]:

Another difficulty associated with economic analysis concerns quantifying economic loss due to damage to a cultural heritage. This task is often impossible. Perhaps this valuation is irrelevant, however, in the same sense as the valuation of a remote wilderness lake or a fish species is irrelevant.

It is to be hoped that the authors do not really mean that such values are "irrelevant." Another view is presented by Glass [46], who states that

Indirect and intangible values in providing recreation, maintaining habitats for wildlife, stabilizing river flows, preventing soil erosion and the siltation of water bodies and aesthetic appearances are all inestimable.

A more encouraging view in this area is provided by Morrison and Sullivan [28] who claim that

Non-commodity values such as wildlife habitat, shelterbelt, watershed protection, setting for outdoor recreation, etc are difficult to estimate but are probably likewise in nine figures.

During the last decade, economists have been developing an analytical framework that will permit them to estimate monetary values for these effects that many people consider "inestimable." The approach is being called a "contingent valuation approach" [47]. In this approach the economist uses a direct personal survey. The interviewer describes a set of possible outcomes for the respondent who is being interviewed. The interviewer then assumes the role of an auctioneer in this constructed contingent market. The interviewer elicits dollar bids from the respondent that serve as a measure of the respondent's "willingness to pay" to have one outcome rather than another.

This approach has been used with a high degree of success in the southwestern region of the United States. Some of these studies have been concerned with the trade-off between measures of environment quality (EQ) and energy developments. As the analytical techniques were being developed the researchers were paying special attention to various biases and sources of errors that could be present in the monetary estimates. These studies have shown that willingness to pay measures obtained through contingent valuation studies can be replicated in repeated contingent valuation studies and can also be replicated with alternative techniques that draw on observed or hypothesized market behavior. As summarized by Schulze et al. [47].

> All evidence obtained to date suggests that the most readily applicable methodologies for evaluating environment quality—hedonic studies of property values or wages, travel cost, and survey techniques—all yield values well within one order of magnitude in accuracy.

A major study commissioned by the Ontario Ministry of the Environment has considered the quantification of these intangible or amenity values associated with the environmental resources susceptible to acidification. The study [48] was conducted by ARA Consultants of Toronto. The study has not yet been released by the ministry, so it is possible to discuss only the survey methods and the theoretical aspects of the approach taken in the study with reference to the questionnaire.

The objectives of the study were

1. to determine the monetary value that members of society place on changes in the quality of environmental amenities caused by acid deposition or other forms of pollution;
2. to determine the socioeconomic factors that might account for the variation in monetary values specified by members of society;
3. to determine the level of awareness regarding acid deposition as well as individual attitudes and beliefs concerning pollution and acid deposition; and
4. to determine the substitution of activities that individuals would make as a result of changes in environmental quality.

The data were collected during a face-to-face personal interview. A total of 920 interviews were conducted; 206 urban residents from the Kitchener–Waterloo area were surveyed in this manner. The remainder were surveyed in "cottage country" which is shown in Figure 3. The major areas were the Muskoka–Haliburton area and the Kawarthas area around Peterborough. In addition to Ontario residents the sample included 100 visitors from the United States.

To obtain monetary estimates of the value of alternative levels of EQ a contin-

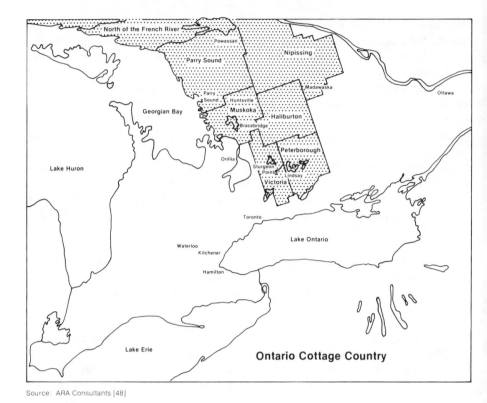

Source: ARA Consultants [48]

Figure 3. Ontario cottage country [48].

gent valuation procedure was employed. To avoid hypothetical bias in the responses, it is necessary to construct scenarios that are believed to be in the realm of possibilities. The respondent was shown an illustrated version of this EQ ladder [48]:

10. "Unpolluted" environment—all fish, wildlife and plants healthy and abundant; restoration of some fish species such as aurora trout and salmon to waters where they used to be naturally, but declined;

9. environment with a wide variety of wildlife, plants, etc.; in some parts of the province the best sports fish no longer exist, such as walleye;

8. Ontario environmental quality level: good fishermen would catch ten fish in two days; Forests, wildflowers and wildlife healthy and in abundance;

7. generally healthy and varied wildlife and vegetation; some decline in sports fish numbers and types (bass); average fishermen would catch seven fish in two days.

6. wildlife and vegetation still relatively healthy and varied; aquatic life not as abundant, fewer frogs and water plants, and loss of some sports fish such as trout; fishermen would catch three fish in two days;

5. wildlife and vegetation declining; fewer otters, ducks and loons; very few fish types; yellow perch, chub and suckers left in lake;
4. only one kind of fish (yellow perch) left; frogs are rare and the small vegetation growth around the lake is rare; water seems clear, and there are very few water birds (ducks, loons) and fish-eating wildlife (raccoons, otters);
3. no fish left in the lake and wildlife that depend on fish for food are gone; water plants except for moss have disappeared and the lakewater is crystal clear except for a green film in the water;
2. the lakewater is very clear, but there is virtually no sign of fish-eating wildlife in the form of ducks or small animals, a very few birds; the trees and shrubs seem thin and the leaves of some are spotted with brown; and
1. polluted environment: fish, wildlife, plants seem less healthy and less abundant; some species no longer exist in the province.

These scenarios were developed with the assistance of scientists at the Ministry of the Environment. The current level of EQ in Ontario is presumed to be at level 8 on the ladder. Changes in the level of EQ was indicated by movements to a different level on the ladder with level 8 as the starting point.

The respondents were shown a schedule of tax allocations (at their income level) for a variety of public services such as defense, police protection, fire protection, education, unemployment insurance and social security. This provides a structure for the respondent to consider his value of the environment by comparing it to other public goods that his taxes currently finance.

The respondent was told that he/she would pay in the form of higher taxes or prices. Instead of the bidding game format used in earlier studies, this study posed an open question to the respondent concerning the amount of money they would be willing to pay to prevent or obtain changes in the level of EQ.

The respondents were asked:

How much, if anything, would you pay in taxes and prices annually to protect the Ontario environment from declining from level 8 to level 4?

The dollar value was recorded and the respondent was asked to explain the reason for the amount stated. This question was repeated three more times with the new level being 7, 2 or 6, respectively. These dollar values are equivalent surplus (ES) measures of consumer surplus for EQ. This dollar value is an amount of income that when given up would result in the same decrease in utility that would occur if the level of environmental quality were to decline. In mathematical shorthand if EQ^o is the initial value of EQ, EQ^1 is the althernative, Y^o is the individual's income level, then the ES value B is determined by

$$U(EQ^o, Y^o - B) = U(EQ^1, Y^o)$$

where U = individual's utility function

It is also possible to obtain a compensating surplus (CS) measure of consumer surplus for environmental quality. This measure can be determined as the amount of income that an individual would be prepared to give up to have an improved environment. He would give up an amount of income that would make him just as well off with the improved environment as he would have been if he had the original level of environmental quality and no change in income. In mathematical shorthand the CS measure B is given by

$$U(EQ^1, Y^o - B) = U(EQ^o, Y^o)$$

To obtain CS measures of consumer surplus for EQ the respondent was asked the following question (after a preamble concerning the ability to improve the level of EQ):

How much, if anything, would you be willing to pay in taxes and prices annually to improve the Ontario environment from 8 to 9?

Again the dollar amount was recorded and the response was probed. This was repeated for a change from 8 to 10. These CS measures are constrained by income and hence should not be subject to problems encountered in earlier studies, which have asked questions concerning compensation for deterioration in EQ. This should permit an analysis of the potential differences between ES and CS measures of consumer's surplus in this study.

To assess the importance of acid rain as a source of damage, the respondents were asked if they would change their bid if the damage had been caused by (1) oil and chemical wastes or (2) natural causes. If the respondents alter their bids, then there exists some information bias. In this case, the monetary estimate may be measuring other effects of acid deposition. The bids would not merely reflect attitudes to changes in EQ in an absolute sense but rather attitudes to acid rain aside from its impact on EQ.

The study will identify both users and nonusers of recreational resources in cottage country. An analysis of the difference between the monetary value for these two groups could shed light on the role of option and existence values of these resources compared to user values.

The study collected data on how respondents would alter their behavior in light of changes in EQ. The respondents were asked questions concerning activity substitution as well as site substitution. The answers to these questions will permit us to perform a travel cost study to compare results with the "willingness to pay" questions. This approach, which is also a contingent valuation procedure, was employed by Thayer [49].

The econometric analysis in the present study is very preliminary, but a further, more detailed formal econometric analysis is anticipated in the near future. It is expected that this study of amenity values will contribute significantly to the understanding of the benefits from the control of acid deposition in Ontario.

CONCLUSIONS

This chapter has surveyed economic theory that may be relevant to a discussion of controlling transboundary pollution. The chapter also reviewed the knowledge concerning the dose-response relationship between acid deposition and physical impacts in the natural environment in order to outline the potential economic significance of acid deposition for eastern Canada.

In some areas such as agriculture and forestry the information concerning the acid deposition dose-response relationship is not sufficiently conclusive to permit estimation of monetary values with confidence at this time.

The aquatic effects of acid deposition are better understood than most other effects. In Ontario alone, 48,000 lakes are considered sensitive to acidification, which could be serious for the sport-fishing industry in Ontario—an important component in Ontario's tourism and recreation industry. In addition to possibly threatening commercial ventures in tourism and recreation as well as those that service this industry, acid deposition may affect intangible services of the environment that members of society enjoy. There is also a possible health risk to people that is associated with the increased uptake of mercury in fish in acidified waters.

I have not attempted to compute a single dollar value associated with damages caused by acid precipitation. Instead, I have presented an array of numbers to indicate potential significance to certain sectors. A single value would suggest a sense of completeness that is inappropriate at this time. The results are not firm enough yet for such a number to be presented. Economic theory and common sense suggest that a cooperative effort in this area between the governments of Canada and the United States will result in higher collective welfare than if each country pursues individualistic policies.

ACKNOWLEDGMENTS

I would like to thank Vante Martini for his assistance. Mr. Martini's assistance was supported by a grant from the Social Science and Humanitites Research Council. In writing this paper I have benefitted from discussions with J.W. Ketcheson, C. Lucyk and D.P. Ormrod. I would like to thank J.A. Donnan and the reviewer for their comments on an earlier draft of this paper.

REFERENCES

1. Asako, K. "Environmental Pollution in an Open Economy," *Econ. Rec.* (1979), pp. 359–367.
2. Pethig, R. "Pollution, Welfare, and Environmental Policy in the Theory of Comparative Advantage," *J. Environ. Econ. Manage.* (1976), pp. 160–169.
3. Markusen, J.R. "Cooperative Control of International Pollution and Common Property Resources," *Quart. J. Econ.* (1975), pp. 618–632.

4. Markusen, J.R. "International Externalities and Optimal Tax Structures," *J. Int. Econ.* (1975), pp. 15–29.

5. Braden, J.B., and D.W. Bromley. "The Economics of Cooperation over Collective Bads," *J. Environ. Econ. Manage.* (1981), pp. 134–150.

6. Forster, B.A. "An Approach to the Optimal Control of Pollution in Boundary Waters," Economics Discussion Paper 78-1, University of Guelph (1978).

7. Forster, B.A. "A Note on Economic Growth and Environmental Quality," *Swed. J. Econ.* (1972), pp. 281–285.

8. Cropper, M.L. "Regulating Activities with Catastrophic Environmental Effects," *J. Environ. Econ. Manage.* (1976), pp. 1–15.

9. Crocker, T.D., and B.A. Forster. "Decision Problems in the Control of Acid Precipitation: Nonconvexities and Irreversibilities," *J. Air Poll. Control Assoc.* (1981), pp. 31–37.

10. "Memorandum of Intent on Transboundary Air Pollution," Group 2 Interim Report, United States–Canada Work Group 2 (1981).

11. Ramsay, F. "A Mathematical Theory of Saving," *Econ. J.* (1928), pp. 543–559.

12. Rawls, J. *A Theory of Justice* (Cambridge, MA: Harvard University Press, 1971).

13. Hartwick, J.M. "Intergenerational Equity and the Investing of Rents from Exhaustible Resources" *Am. Econ. Rev.* (1977), pp. 972–974.

14. Asako, K. "Economic Growth and Environmental Pollution under the Max-Min Principle," *J. Environ. Econ. Manage.* (1980), pp. 157–183.

15. Rubec, C.D.A. "Characteristics of Terrestrial Ecosystems Impinged by Acid Precipitation Across Canada," Working Paper 19, Lands Directorate, Environment Canada Ottawa, Ontario (1981).

16. "Memorandum of Intent on Transboundary Air Pollution," Interim Report, United States–Canada Work Group 1 (1981).

17. Linzon, S.N., R.G. Pearson, W.I. Gizyn and M.A. Griffith. "Terrestrial Effects of Long Range Pollutants—Crops and Soils," Proceedings of the APCA Conference, Acid Deposition, Knowns and Unknowns: The Canadian Perspective, Montreal, Quebec, April 7–8, 1981.

18. "Memorandum of Intent on Transboundary Air Pollution," Phase II Report, United States–Canada Work Group 1 (1981).

19. Harcourt, S.A., and J.F. Farrar. "Some Effects of Simulated Acid Rain on the Growth of Barley and Radish," *Environ. Poll.* Ser. A, 1980, pp. 71–72.

20. Ketcheson, J.W. "Long-Range Effects of Intensive Cultivation and Monoculture on the Quality of Southern Ontario Soils," *Can. J. Soil Sci.* (1980), pp. 403–410.

21. Ormrod, D.P., J.T.A. Proctor, G. Hofstra and M.L. Phillips. "Air Pollution Effects on Agricultural Crops in Ontario: A Review," *Can. J. Plant Sci.* (1980), pp. 1023–1030.

22. "Agricultural Statistics for Ontario 1980," Ontario Ministry of Agriculture and Food, Toronto, Ontario (1981).

23. "The Acid Earth," *Harrowsmith* (1980), pp. 32–93.

24. Ketcheson, J.W. Personal communication (1982).

25. Underwood, J.K. "Acidic Precipitation in Nova Scotia," Department of the Environment, Halifax, Nova Scotia (1981).

26. MacLean, R.A.N. "Statement on Terrestrial and Aquatic Effects," in Proceedings the APCA Conference, Acid Deposition, Knowns and Unknowns: The Canadian Perspective, Montreal, Quebec, April 7–8, 1981.

27. Statistics Canada. *Canada Year Book* (Ottawa, Ontario: Queen's Printer, 1980).

28. Morrison, I.K., and C.R. Sullivan. "Terrestrial Effects—Forest Ecosystems," Proceedings of the APCA Conference, Acid Deposition, Knowns and Unknowns: The Canadian Perspective, Montreal, Quebec, April 7–8, 1981.

29. "The Long Range Transport of Air Pollutants," Second Report, United States–Canada Research Consultation Group (1980).

30. Brydges, T. "Aquatic Effects of Acid Deposition—Water Quality," Proceedings of the APCA Conference, Acid Deposition, Knowns and Unknowns: The Canadian Perspective, Montreal, Quebec, April 7–8, 1981.

31. Haines, T. "Acidic Precipitation and Its Consequences for Aquatic Ecosystems: A Review," *Trans. Am. Fish. Soc.* (1981), pp. 669–707.

32. Beamish, R.J. "Acidification of Lakes in Canada by Acid Precipitation and the Resulting Effects on Fishes," *Water, Air Soil Poll.* (1976), pp. 501–514.

33. Cooley, J.M. "Aquatic Effects: Water Quality and Fisheries," Proceedings of the APCA Conference, Acid Deposition, Known and Unknowns: The Canadian Perspective, Montreal, Quebec, April 7–8, 1981.

34. Beamish, R.J., and H.H. Harvey. "Acidification of the LaCloche Mountain Lakes, Ontario, and Resulting Fish Mortalities," *J. Fish. Res. Board Can.* (1972), pp. 1131–1143.

35. Beamish, R.J. "Loss of Fish Populations from Unexploited Remote Lakes in Ontario, Canada as a Consequence of Atmospheric Fallout of Acid," *Water Res.* (1974), pp. 85–95.

36. "The Case Against the Rain," Ontario Ministry of the Environment, Toronto, Ontario (1980).

37. Statistics 1980, Ontario Ministry of Natural Resources, Toronto, Ontario (1981).

38. "Final Report on Acidic Precipitation, Abatement of Emissions from the International Nickel Company Operations at Sudbury, Pollution Control in the Pulp and Paper Industry, and Pollution Abatement at the Reed Paper Mill in Dryden," Legislature of the Province of Ontario, Standing Committee on Resources Development (1979).

39. Lucyk, C. "Ontario's Program of Socio-economic Studies," Proceedings of the APCA Conference, Acid Deposition, Knowns and Unknowns: The Canadian Perspective, Montreal, Quebec, April 7–8, 1981.

40. "Statistics on Sales of Sport-Fishing Licences in Canada, 1972–1977," Fisheries and Oceans, Ottawa, Ontario.

41. Liddle, G. "Potential Socio-economic Impacts of Acid Rain on the Fishing Lodge Industry of Northern Ontario," Proceedings of the Action Seminar on Acid Precipitation, November 1–3, 1979.

42. Donnan J.A. "Ontario Socio-economic Research Program," paper presented at the Canadian Sport Fisheries Conference, Calgary, Alberta, July 19, 1981.

43. Bengtsson, B., W. Dickson and P. Nyberg. "Liming Acid Lakes in Sweden," *Ambio* (1980), pp. 34–36.

44. Lucyk, C. Personal communication (1982).

45. "The Acid Lakes," *Globe and Mail,* Toronto, Ontario (July 15, 1980).

46. Glass, N. "Effects of Acid Precipitation," *Environ. Sci. Technol.* (1979), p. 1352.

47. Schulze, W.D., R.C. d'Arge and D.S. Brookshire. "Valuing Environmental Commodities: Some Recent Experiments," *Land Econ.* (1981), pp. 151–172.

48. "The Effects of Acid Deposition on Environmental Amenities," ARA Consultants, report to the Ontario Ministry of the Environment (1982).

49. Thayer, M. "Contingent Valuation Techniques for Assessing Environmental Impact: Further Evidence," *J. Environ. Econ. Manage.* (1981), pp. 27–44.

CHAPTER 8

Legal, Ethical, Economic and Political Aspects of Transfrontier Pollution

Allen V. Kneese
Ralph C. d'Arge

State responsibility and liability are not clearly defined with respect to cases of transboundary pollution, such as environmental degradation by acid rain in one state resulting from emissions in another. However, numerous declarations by international agencies and court opinions do exist. For example, the Organization for Economic Co-operation and Development (OECD) countries unanimously subscribed to the 1972 United Nations (UN) Declarations on the Human Environment, particularly to Principle 21, which states:

> States have, in accordance with the Charter of the United Nations and the principles of international law, the sovereign right to exploit their own resources pursuant to their own environmental policies, and the responsibility to ensure that activities within their jurisdiction or control do not cause damage to the environment of other States or of areas beyond the limits of national jurisdiction [1].

This is an apparently clear cut statement that, on its face, would simply make transfrontier pollution, which caused damages, inadmissable. However, in a later OECD document [2], this principle is weakened considerably.

> Countries should endeavour to prevent any increase in transfrontier pollution, including that stemming from new or additional substances and activities, and to reduce, and as far as possible eliminate, any transfrontier pollution existing between them within time limits to be specified [2].

This statement recognizes transfrontier pollution as a fact but proscribes any increase in it and requires efforts to reduce it.

The leading international arbital award in the matter of transfrontier pollution is the Trail Smelter Case [3]. It and other evidence has led a leading legal expert to describe the pertinent rule in international law as follows:

It is unlawful for a state to cause transfrontier pollution which entails *serious* [emphasis added] damage in another state. The rule can also be reversed: Pollution below the threshold of seriousness is to be tolerated. If transfrontier pollution is above the threshold of "seriousness" there is not only a duty to pay damages, but to prevent future injury, to refrain from further polluting the neighbouring territory [1].

In yet another pronouncement on responsibility for pollution, the OECD adopted the "polluter pays" principle in 1972. This states that the polluter must pay for any ameliorating measures that are caused to be undertaken [2]. The principle does not, however, require payment of compensation for any residual damage that may remain. It also does not specify how the level or intensity of pollution reduction is decided on. However, a German law that has recently gone into effect requires payment of an effluent charge for any wastewater discharge that presumably, to some extent, is meant to reflect remaining damages within German territory.

Although not all of these statements are fully consistent with each other, the tendency clearly is to interpret state responsibility as requiring that states within whose boundaries harmful actions occur must pay, or cause to be paid by responsible parties, the cost of ameliorating those actions and, in some interpretations, the remaining damage as well.

EFFICIENCY

d'Arge and Kneese defined and interpreted in economic terms several major principles for assigning state responsibility in transboundary pollution cases [4]. The first principle is that each state is responsible for all emission control internally and externally but is not responsible for compensation of remaining damages following installation of the agreed-on controls. This is a variant of the OECD polluter-pays principle cited above in that it is applied to transfrontier pollution problems; application of the principle to such problems was explicitly excluded by OECD member nations.

The second major principle is the full costing principle (FC). This requires the state responsible for waste discharge to pay compensation for remaining damages as well as control costs and would be economically efficient, since the polluting country would be comparing the cost of control with payment for compensation. However, implementation of such a principle requires the state to accept international responsibility, which may reduce national sovereignty and perhaps, national wealth.

The third principle is the "victim pays" principle (VP). This requires the affected state to compensate the affecting state (or internal parties creating harmful residuals) for costs of control and to absorb residual damages after controls are implemented. We include the VP principle not because it appears to have any inherent ethical appeal but because in some instances it may be the only practical possibility. As a real example, recently a tentative agreement was reached under which the French will reduce pollution of the Rhine with salts from Alsatian potash mines. However, the agreement was reached only after years of haggling and only after the downstream countries agreed to make a large contribution toward meeting the costs of pollution control [5].

After extensive analysis of various situations, d'Arge and Kneese concluded as follows:

> We have attempted to demonstrate that there is no overarching principle of state responsibility such as "the polluter pays" which is politically or economically (in the efficiency sense) applicable to the entire spectrum of transnational environmental problems. In bilateral cases of transnational pollution, the adoption of a full costing principle by all nations appears to be relatively efficient if the emitter country does not compensate the receptor or the receptor country does not utilize the payments by the emitter country to compensate firms (or individuals) for damages. In multilateral cases, the FC principle also appears to be the most efficient if the number of receptor countries is large relative to the number of emitters, because negotiation costs can be expected to be lower. It should be noted that in both cases, if negotiation or other transactions costs are significant, a viable internationally agreed-upon rule for identifying who must incur these costs and setting penalties for not doing so needs to be specified [4].

Thus, it appears that the polluter-pays plus compensation (to the country, not the specific receptors) rule in many instances has desirable efficiency properties. But this is only part of the story, perhaps only a small part of it, because:

1. An outcome based on FC may not accord with beliefs held about equity and justice.
2. Since there is no clear international enforcement authority, real outcomes are determined by international political forces.
3. For some cases, like acid rain and carbon dioxide pollution of the stratosphere, countries are both emitters and receptors.

In addition, there is the very important problem of whether sufficient information exists to implement the FC (or for that matter any) principle in dealing with transfrontier pollution. Since this chapter is primarily concerned with principles and concepts, we will set aside the information question and provide a brief discussion of the other two issues.

JUSTICE AND EQUITY

The various pronouncements of principles concerning transboundary pollution are probably based as much on ideas about equity and justice as on efficiency. Moral philosophers have long concerned themselves with such ideas, especially with respect to the behavior of individuals. Schulze and Kneese tried to apply some highly specialized principles of moral philosophy to some large societal decision problems [6]. Here we select three of these principles, first discussing them as they pertain to individual behavior and then attempting to generalize to the international context.

A utilitarian ethical system requires "the greatest good for the greatest number" as expressed in the eighteenth and nineteenth centuries by Jeremy Bentham [7] John Mill [8] and others. Effectively, the social objective of this criterion is to maximize the sum of the cardinal (measurable) utilities of all individuals in a society. Thus, for an individual person to act in an ethically "correct" way, all consequences of that action must be considered. Thus, the utilitarian ethic has a pragmatic consequentialist character which, in a matter-of-fact way, is quite appealing.

However, in addition to the obvious difficulty in making all of the requisite calculations necessary for moral choices, a fundamental problem afflicts utilitarian theory when applied to social decision-making—measuring utility. The matter of distributing income will serve to demonstrate the problem of measurable or cardinal utility. First, we will make the assumption (consistent, for example, with the view of Pigou [9] that all individuals have about the same relationship between utility (U_A and U_B, respectively) and income (Y_A and Y_B, respectively). If Mr. B is initially wealthier than Mr. A, $Y_B^0 > Y_A^0$, then B has a higher total utility level than A. However, given the traditional utilitarian assumption of diminishing marginal utility, it is easy to show that society's total utility could be enlarged by giving A and B the same income $\bar{Y}$. This follows because, by raising A's income from Y_A^0 to $\bar{Y}$, we get a gain in utility of $\triangle U_A$ to A compared to the loss in utility $\triangle U_B$ to B, resulting from lowering B's income from Y_B^0 to $\bar{Y}$. Note that $Y_B^0 - \bar{Y} = \bar{Y} - Y_A^0$, so we take income away from B to give to A to get a gain in total utility, $U_A + U_B$, since $|\triangle U_A| > |\triangle U_B|$, or A's gain exceed B's loss.

The same solution results from solving the following problem:

$$\text{maximize} \quad U_A(Y_A) + U_B(Y_B)$$
$$\text{subject to} \quad Y_A + Y_B = Y_A^0 + Y_B^0$$

which implies at the optimum that $\triangle U_A / \triangle Y_A = \triangle U_A / \triangle Y_B$, or that the rate of increase of utility with income (marginal utility) must be equal for the two individuals. Since the two individuals in our example have similar utility functions, marginal utilities are equated where incomes are the same, $Y_A = Y_B = \bar{Y}$.

On the other hand, we can assume different individuals have different utility functions. For example, Edgeworth [10] argues that the rich have more sensitivity and can better enjoy money income than the poor. Mr. A then gets more income than Mr. B because he obtains more utility from income that B does. In Edgeworth's view, Mr. A by his sensitivity should have more money to be used in appreciating fine wine than Mr. B, who is satisfied with common ale.

Clearly, then, depending on beliefs about the particular nature of utility functions, any distribution of income can be justified, ranging from a *relatively* egalitarian viewpoint (Pigou) to a *relatively* elitist viewpoint (Edgeworth).

There do exist ethical ideas that are totally egalitarian and totally elitist. We realize that probably very few people, if pushed to the wall, would actually support either of these extreme systems in its pure form. But it is useful to analyze at least the egalitarian system, which has great appeal to many people, as representing an end of the spectrum.

The egalitarian view holds that the wellbeing of a society is measured by the wellbeing of the worst off person in that society. This criterion would lead, if fully adopted, to a totally equal distribution of utility [11]. We must add that Rawl's theory of just distribution is much more complex than the simple egalitarian criterion we analyze here.)

The egalitarian criterion can be expressed in symbolic form as follows: for two individuals A and B, where utility is denoted U, if $U_A < U_B$, we maximize U_A subject to $U_A \leq U_B$; if $U_B < U_A$, then we maximize U subject to $U_A \geq U_B$. If we reach a state where $U_A = U_B$, then we maximize U_A subject to $U_A = U_B$. The implication of this for redistribution of income is that we begin by adding income to the worst off individual (taking income away from wealthier individuals) until he catches up with the next worst off individual. We then add income to both individuals until their utility levels (wellbeing) have caught up to the third worst off, etc. Eventually, this process must lead to a state where $U_A = U_B = U_C = U_D$ . . . for all individuals in a society, where all utilities are identical, or to one where further redistributions will make everyone worse off, e.g., through negative impacts on incentives.

The third ethical system is an amalgam of a number of ethical principles embodied in part in a Christian ethic (the Golden Rule) as well as in the U.S. Constitutional viewpoint that individual freedoms prevail except where others may be harmed. These views, which emphasize individual rights, have been formalized by Nozick [12] in a strict libertarian framework. We are not concerned here with changing the initital position of individuals in society to some ideal state, as were the ethical systems discussed earlier, but rather in benefiting all or at least preventing harm to others, even if those others are already better off. This ethic has been embodied often by economists in the form of a criterion requiring "Pareto superiority," that is, an unambiguous improvement in welfare requires that all persons be made better off by a change in resource use or at least as well off as before. Any act is then immoral or wrong if anyone is worse off because of it. Any act which improves an individual's or several individuals' wellbeing and harms no one is then moral or "right."

If, for example, Mr. A and Mr. B initially have incomes Y_A^0 and Y_B^0, then we require for any new distribution of wealth (Y_A, Y_B)—for example, if more wealth becomes available and must be distributed—that

$$U_A(Y_A) \geq U(Y_A^0)$$

and

$$U_B(Y_B) \geq U(Y_B^0)$$

or each individual must be at least as well off as he initially was. Any redistribution, e.g., from wealthy to poor or vice versa, is specifically proscribed by this criterion.

Thus, this criterion, while seemingly weak, i.e., it does not call for redistribution, can block many possible actions if they do as a side effect redistribute income to make *anyone* worse off, however slight the effect may be. Often, then, to satisfy a libertarian criterion requires that gainers from a particular social decision must *actually* compensate losers.

To think of these alternative rules for moral behavior in the international context of transboundary pollution may seem farfetched. However, it should be noted that relations between sovereign nations present many of the same ethical problems as relations between individuals and that the rules of international behavior discussed in the previous section bear a strong family resemblance to the ethical principles just reviewed. For example, article 21 of the UN Declaration of the Human Environment is essentially a libertarian rule, with respect to nations, which instructs one nation not to reduce another nation's welfare *at all* by imposing pollution on it. What we have termed the full costing rule is also libertarian in character by admitting the possibility of compensation to nations (but not specific receptors). The Trail Smelter rule could be interpreted as a utilitarian rule that permits a state to impose some damage on another to permit the former to realize utility from its resources. However, it must cease doing so if the damage is serious (the utility lost outweighs the utility gained?). We are not aware of a principle of transboundary pollution that corresponds to the egalitarian rule although many persons would no doubt hold that a "rich" nation polluting a "poor" nation is somehow "worse" than the reverse.

To carry these ideas just a bit further let us assume that we can think of a nation as having a "utility" function (some aggregation of the utility function of all its citizens). Let us further assume that at issue is transboundary pollution between two nations, A and B. Even further let us assume that the utility functions of the two countries are the same (without stretching the point too much it could be argued that this is not an incredible description of Canada and the United States). So that we can address a situation rigorously suppose further that country A pollutes country B, that compensation is not an option, and that the marginal loss in monetary terms, in country B is exactly equal to the incremental gain in country A. Results of an analysis of this situation is presented in Table I. Y is used to symbolyze income [6].

Table I. Ethical Views of an International Pollution Problem: A's Prospective Loss Equal to B's Current Damage

Ethic	$Y_A^0 > Y_B^0$	$Y_A^0 < Y_B^0$
Utilitarian	Reject	Accept
Egalitarian	Reject	Accept
Libertarian	Reject	Reject
Traditional B/C	Accept	Accept

The example is structured so that standard cost/benefit analysis just accepts the imposition of an increment in uncompensated cost. However, the utilitarian ethic rejects the situation if country B is worse off than A, i.e., B has a lower per capita income, $Y_B^0 > Y_B^0$. (If we were considering Canada and the United States income distribution would probably carry little weight—per capita income in the two countries is about equal and any credible pollution situation seems unlikely to change that situation much.) Similarly, the egalitarian criterion rejects the imposition of risk by country A on country B if B is worse off initially. (Again, this criterion would have little bearing.) Finally, the libertarian ethic rejects the notion of uncompensated risk no matter what the initial distribution of wealth since, by definition, an uncompensated cost makes one nation worse off.

Thus, if we want to apply these ideas roughly to the situations between Canada and the United States, and if the distribution of costs and gains is as indicated, the utilitarian criterion (aimed at maximizing the utility of the entire group at issue) would accept the pollution, the libertarian criterion (aimed at protecting individual rights) would reject it, and the egalitarian criterion would be largely irrelevant. Finally, if compensation to the disadvantaged nation were possible (the FC principle) arrangements could be made which none of the ethical systems found offensive and that at the same time were relatively efficient. This is one of the appealing aspects of the FC principle, since explicit ethical beliefs do not have to be evaluated and agreed on in the international political arena. Also, in Canada and the United States there is likely to be a cross section of ethics, and then the critical issue arises for decision-making as to how to weigh them.

We should be clear that we are not trying here to make a judgment about which criterion is "best." Rather, we are trying to make more explicit the implications of some widely held ethical views that underlie, or at least can be read into, certain international declarations about transboundary pollution.

TRANSFRONTIER POLLUTION, INTERNATIONAL POLITICS, AND THE DISTRIBUTION OF COSTS AND GAINS

Regardless of the possible desirability, in principle, of applying criteria based on ideas concerning efficiency or justice to transboundary pollution, the absence of an international enforcement authority and the dominance of national self-interest may make these largely irrelevant to actual outcomes. In a two party upstream-downstream case (whether air or water) there is a well defined victim and a well defined "causer." In such a case if the victim has no bargaining power, the VP principle will be the operative one as appears to be the case with respect to the Netherlands and the Rhine noted earlier. In the case where there is room for bargaining, both parties will likely pay but not necessarily in the same currency.

To illustrate this point it will be useful to look in a little detail at another transfrontier pollution problem between the United States and a neighbor, in this case Mexico. The problem in this instance is salinity levels in the Colorado River water delivered across the border to Mexico.

A treaty was signed between Mexico and the United States in 1944 dividing up the water of the Colorado between the two nations. It said nothing about water quality. For most of the period since 1944, the water delivered to Mexico, stored in Morelos Reservoir, was not much worse than that delivered to users in the lower basin in the United States. Nevertheless, the gradual reduction of water quality due to increased use in the United States would most probably have caused salinity to arise as an international issue in due course.

In the early 1960s, a dramatic fall in the quality of water being delivered to Mexico occurred. In 1947 the Bureau of Reclamations' Wellton–Mohawk Project in southwestern Arizona was authorized by Congress to deliver water for the irrigation of 75,000 acres. As a solution to salinity problems the project encountered, in 1961 the Wellton–Mohawk irrigation district started pumping from drainage wells and discharging saline water into the Colorado River below the last U.S. diversion point, but above the Mexican diversion point. This water with a salinity of about 6000 ppm caused a sharp increase in the salinity of the water delivered to Mexico—from about 800 ppm in 1960 to more than 1500 ppm in 1962. The effect of the increased salinity discharge was reinforced when the quantity of water delivered to Mexico dropped sharply because, in 1961, Glen Canyon Dam commenced filling. Mexico complained loudly.

Following some interim activities to try to blunt the problem, in August of 1972, President Nixon appointed Herbert Brownell, Jr., as his special representative on the problem with Mexico and assigned him to the job of finding a "permanent" solution to the Mexican salinity issue. After delivering a report on the matter, Brownell was appointed special ambassador to negotiate a solution with Mexico. The result of that negotiation was incorporated in Minute 242 of the International Water and Boundary Commission. This minute refers to itself as the "permanent and definitive solution" to the international salinity problem.

The Administration pledged itself to undertake the following measures:

1. construction of a major desalting plant and related works for Wellton–Mohawk drainage waters;
2. extension of the Wellton–Mohawk drain (for the brine from the plant) to the Gulf of California;
3. lining or construction of a new Coachella Canal in California; and
4. improved Wellton–Mohawk irrigation efficiency.

In a situation like this, if one views the matter from a narrow economic point of view, there can be no basis for agreement unless the victim country compensates the damaging country in costs incurred to reduce the damage (VP principle). If the damaging country acts in its own, narrowly defined, immediate economic self-interest, it will not be willing to reach any other sort of agreement.

In the case of the Colorado, the United States agreed to pay the entire cost of mitigating the situation. If the United States had been acting on the basis of a narrow interpretation of economic self-interest, it either would have done nothing or would have required Mexico to pay the costs of mitigation. This, then, leads one to wonder what factors, other than magnitudes and distribution of strictly economic costs and benefits or more broadly, economic self-interest, could be important in the U.S. decision. Economic self-interest is, of course, always pertinent to the development of national decisions. But if it had been the only consideration, the situation never would have turned out the way it did. Since the United States paid for the control costs but did not compensate for the remaining damages to Mexico, this is an example of the "polluter pays" principle.

It appears that in many cases negotiations about transfrontier pollution or other external effects involve considerations that are in quite other arenas of national interest, and in many cases these "extraneous" considerations have been dominant. A much wider view of the national and international considerations involved must be taken. Indeed, it may be that even the initiation of negotiations on an international externality problem is more related to other considerations of national interest— and clearly the results often are. For example, Krutilla has shown that various considerations, including military strategic ones, caused the United States to agree to a treaty on the Columbia which was quite unfavorable to it economically [13]. (The Columbia situation involved no pollution problem but rather the possibility of both nations benefiting from joint rather than individual development. This is because it was possible in this way to benefit from positive externalities. Thus, generically it is like a transfrontier pollution problem. However, in this case, it is in the interest of both parties to negotiate a deal.) Thus, matters such as trade concessions, military bases and the desire to win allies in international politics may often be overriding considerations.

In addition, the image a country wishes to project to the outside world may sometimes be important. One might interpret the agreement of the Swiss, who are at the head of the Rhine River, to pay 5% of the costs of reducing salt discharges in France as being a result of this objective. A related consideration might be the attitude of a particular country toward international law. Our own conjecture is that these latter considerations are not nearly so important as the more direct national self-interest-related ones mentioned earlier. In the case of international conflicts between states, as long as all interested parties have something of value to trade, the outcome of this kind of process may not be so arbitrary or irrational as it might at first seem—even though the outcome will most likely be in conflict with some efficiency and/or equity criteria. Indeed, as long as national self-interest is the primary motivating force in international affairs, as we believe it is, this type of broad trading process seems to be the only one that can dependably lead to international agreements, especially in the type of situation that characterized the salinity problem of the Colorado River.

Actually, such larger factors seem to have been very important in all the major Mexican–American water agreements that have been achieved. During the 1939–1944 negotiations over the Colorado and the Rio Grande, one factor that caused the United States to agree to an allocation of 1.5 million ac-ft per year of Colorado River water to Mexico, an amount far above historical usage in the

Mexicali Valley, was its desire to cultivate friendly relations with other countries, especially neighboring countries, during World War II. In that case, however, the situation was more complex than the salinity situations, since agreements for both rivers were being negotiated simultaneously, and the United States *did* have a direct economic interest in reaching an agreement on the Rio Grande. This was so since most of the water in the lower Rio Grande (below Fort Quitman, Texas) originates from the Mexican tributaries. For this reason, the Mexicans insisted on negotiating about both rivers at the *same* time.

The 1973 agreement also seems to have been a situation in which the United States was anxious to cultivate more favorable relations with Latin American countries and, more generally, to reduce international stresses in the world at large. More particularly, the United States already had stresses along the Mexican border, for example, having to do with illegal immigrants, that it did not wish to exacerbate.

CONCLUSIONS

In this chapter, both compensated and uncompensated transnational pollution problems have been analyzed. Compensation among nations is necessary to achieve economic efficiency, in that costs and benefits will be adequately compared. For many Western ethical criteria, uncompensated risks imposed by one nation on others could be unethical, but this depends on the ethical criterion as well as differences in income and other political considerations. Because of differences in the mix of ethical beliefs among nations along with the desire to maintain national sovereignty, no all-inclusive principle is likely to emerge to solve transnational pollution problems. What is likely to emerge are a sequence of relatively unique solutions that blend economic necessity with political cunning and goodwill. However, the first step appears to be that every nation agrees to negotiate in good faith toward a solution of transnational pollution provided there is sufficient evidence of economic benefit and cost.

The record on international settlement of transnational pollution does not lead to a clear and concise set of guiding international principles. The various cases demonstrate the various possibilities ranging from full costing to victim pays. It is unlikely that such piecemeal solutions will lead to economic efficiency in a global sense. Yet, as long as there are gains from trade, including pollution, among nations, and these gains from trade are negotiated over and obtained, everyone should be as well off or better than without negotiation.

REFERENCES

1. "Transfrontier Pollution and the Role of the States," Organization for Economic Co-operation and Development, Paris (1982).
2. "Principles Concerning Transfrontier Pollution." [(74)224)], paragraph 3.
3. International Arb. Awards 1905, 1963, Trial Smelter Case (United States vs. Canada), 3R (1935).

4. d'Arge, R.D., and A.V. Kneese. "State Liability for International Environmental Degradation: An Economic Perspective," *Natural Resources J.* 20(3):427–450 (1980).
5. *Economist* (November 28, 1981).
6. Schulze, W.D., and A.V. Kneese. "Risk in Benefit-Cost Analysis," *Risk Analysis* 1(1):81–88 (1981).
7. Bentham, J. *An Introduction to the Principles of Morals and Legislation* (Oxford, Clarendon, 1789).
8. Mill, J. *Utilitarianism* (New York: Longmans, Greene & Company, 1863).
9. Pigou, A.C. *The Economics of Welfare* (London: Macmillan & Company, 1920).
10. Edgeworth, F. *Mathematical Psychics: An Essay on the Application of Mathematics to the Moral Sciences* (New York: A.M. Kelley, 1967).
11. Rawls, J. *A Theory of Justice* (Cambridge: Belknap Press, 1971).
12. Nozick, *Anarchy, State, and Utopia* (New York: Basic Books, 1974).
13. Krutilla, J.V. "The International Columbia River Treaty: An Economic Evaluation" in *Water Research,* A.V. Kneese and S.C. Smith, Eds. (Baltimore, MO: John Hopkins University Press, 1965).

CHAPTER 9

Acidification Impact on Fisheries: Substitution and the Valuation of Recreation Resources

Fredric C. Menz

John K. Mullen

Acid precipitation has emerged in recent years as a major environmental issue, possibly as a result of increasing concern about its impact on natural ecosystems and human health and welfare. Although there has been considerable research into scientific aspects of the acidification problem, major gaps in understanding the issue remain. Research into economic aspects of the problem has been relatively limited, despite the importance of economic considerations at the policymaking level. The implications of the acid deposition issue for national energy policy may be profound, so the role of economic analysis in examining the problem should be understood.

Several features distinguish acid deposition from other environmental pollution problems, making the design of policy particularly difficult. One factor of importance is the substantial uncertainty concerning the magnitude of both the short- and longer-term effects of acid deposition. Problems associated with uncertainty may be especially important in this case because some of the impacts from acidification may be irreversible. In addition, certain types of acidification damage may be characterized by thresholds rather than being a continuous function of acidic inputs. Other unique features include the widespread geographic area potentially affected by acid precipitation and possible discrepancies between public perception of the issue and its actual impacts. These factors suggest that conventional cost-benefit analysis may lead to inappropriate policy recommendations for the control of acid-forming emissions.

This chapter presents the results of research concerned with estimating the economic significance of damages due to increased acidity in the Adirondack recreational fishery. The nature of the acid deposition problem and the characteristics of the Adirondack fishery present a number of significant conceptual and empirical problems in attempting to assess the economic magnitude of damages from increased acidity. This chapter is particularly concerned with the appropriate treatment of substitution possibilities in an economic framework for valuing acidification damages. The issue involves identifying and considering relevant substitution possibilities for estimating the value of the Adirondack recreational fishery and also for determining the loss from acid precipitation. While the magnitude of damage to the freshwater ecosystem is likely to exceed that resulting from damage to the recreational fishery alone, damages associated with a decline in the quality of the fishery are likely to be an important component of the overall loss in economic welfare. Thus, the results of this empirical research provide a lower-bound estimate of the economic damages from acidification in the Adirondacks.

BACKGROUND

Scientific Aspects

Detailed discussion of the physical and chemical processes that cause acid deposition is beyond the scope of this chapter. For examination of the scientific processes involved in the formation and long range transport of acid precipitation, see the literature [1–6]. Acid deposition can potentially affect arable land, forests and aquatic ecosystems [7–10].

The gradual accumulation of acidic components, whether from acid precipitation or from other sources, can affect aquatic ecosystems at all levels, but the effects on fish are possibly the best understood and most widely documented [11,12]. Acidification of lakes and rivers accompanied by loss of fish populations has been observed in Scandinavia [13], the Adirondack Mountains [14,15] and southeastern Ontario [16,17]. Factors common to these areas include certain predisposing geological and morphological conditions, soft and poorly buffered surface waters, and decidedly acidic precipitation [18–20]. Except for the Sudbury region of Ontario, the affected areas do not have significant local emissions capable of contributing substantially to the problem.

While the mechanisms of acidic input into freshwater bodies have been recognized, the relative importance of acid precipitation as a contributing factor has not been resolved. There have been numerous reports of the depletion of fish stocks in dilute waters exposed to acid precipitation, but the mechanisms leading to the loss of fish are not thoroughly understood. Increased acidity alone affects fish reproduction, but the combination of acids and aluminum released by the acids has been identified as the primary factor limiting fish survival in Adirondack waters [21,22].

Adirondack Fishery

The Adirondack fishery is comprised of approximately 3000 individual lakes and ponds, encompassing some 282,154 surface acres and about 32,000 miles of streams and rivers (Table I). About 80% of the total water acreage is open to public fishing. The boundary of the fishery is nearly coincidental with that of the Adirondack Park, which is comprised of 6 million acres of relatively undeveloped and sparsely populated land within a day's traveling distance of an estimated 50 million residents of the United States and Canada. A comprehensive survey in 1975 found a surface pH of <5.0 in 52% of 217 high elevation ponds, and 82% of the acidified waters were devoid of fish [15]. Comparable data suggest that only 4% of these waters were similarly acidified and devoid of fish in 1927–1937, and that decreases in pH have occurred over a wide range of pH values for a subsample of Adirondack waters since that time [14,15].

According to the most recent inventory of the acidity status of Adirondack region waters, 25% of the 849 waters surveyed since 1974, comprising 11,000 acres, have reached a critical state of acidification, exhibiting a pH of <5.0 (Table II). These acidified waters represent about 10% of the region's lakes and nearly 4% of total Adirondack Zone water acreage. Nearly one-half of the lost acreage involves remote former brook trout ponds. The impact on brook trout has been

Table I. Inventory of Waters in the Adirondack Region

Classes of Waters[a]	Total[b]		Percentage of Total Open to Public Fishing[c]	
	Number	Acreage	Number	Acreage
Brook Trout Ponds	512	28,407	59	53
Two-Story Ponds/Lakes	65	102,163	84	97
Coldwater Ponds/Lakes	72	5,655	85	89
Warmwater Ponds/Lakes	437	112,689	74	83
Unknown Ponds/Lakes	1,527	21,541	66	54
Known Acidified Ponds/Lakes	264	11,518	81	74 .
Totals	2,877	282,154	67	81

Significant Streams	Total[b]		Percentage of Total Open to Public Fishing[c]	
	Miles	Acreage	Miles	Acreage
Coldwater Streams/Rivers	5,097	9,844	69	69
Warmwater Streams/Rivers	708	6,852	67	67
Smaller Tributary Streams	26,000	NA[d]	NA	NA
Totals	31,805	16,696	69	68

[a] Water classes are defined as follows: brook trout: managed primarily for brook trout; coldwater: managed for "several" salmonid species; warmwater: managed primarily for warmwater species; known acidified: includes 16 limed ponds (currently viable), 36 known naturally acid ponds and 212 ponds acidified post-1974; two-story: managed for fishable populations of both coldwater and warmwater species.
[b] From Pfeiffer and Festa [14].
[c] Computed from data in Pfeiffer [23].
[d] NA = data not available.

Table II.   Overall Acidity Loss Based on Subsample of Ponded Waters for Which Post-1974 Surface pH Data are Available, 1980 [14]

pH Range	Status	Acreage		Number of Waters	
		Total	Percentage	Total	Percentage
<5.0	Critical	10,460	4.7	212	25.
5.0–6.0	Endangered	63,243	28.4	256	30.1
>6.0	Satisfactory	149,021	66.9	381	44.9
Total		222,724		849	

greatest not because they are particularly sensitive to increased acidification (brook trout are one of the more acid-tolerant salmonid species), but because they were the only resident game fish species in many of the affected ponds. Brook trout are a popular native game fish in the Adirondacks, and many of the affected brook trout waters provided the potential for a genuine wilderness fishing experience [23]. Attempts to reestablish brook trout populations by stocking affected waters have not been particularly successful. Current efforts to restore or preserve fish stocks in acidified waters center on liming to neutralize acidity [24]; however, careful monitoring is required to avoid increased heavy metal toxicity that frequently occurs with liming. Another area of current research involves selective breeding of fish strains with increased tolerance to acidified water and its associated chemistry [12,25].

Initial concern with acidification in the Adirondacks emphasized the loss of habitat for coldwater species, particularly brook trout ponds, but acidification appears to be endangering warmwater habitats in the Adirondack region as well [14]. A number of former bass lakes in the Adirondack region, including Woodhull Lake, Big Moose Lake and Canada Lake, have lost their smallmouth bass populations in the recent past, and reproductive failure associated with increased acidity is suspect. Population declines are also suspected in several other large bass waters with a marginal pH, including Cranberry Lake and Tupper Lake, but the matter is still under investigation [26]. A recent survey of Adirondack streams found 15% of sampled streams to be in the critical state [27]. Aquatic life in streams is particularly vulnerable to the severe pH fluctuations that accompany spring melt of the winter snow cover.

VALUING ENVIRONMENTAL DAMAGES

Economic Approach

Several methods have been used to estimate the value of damages from environmental pollution [28]. Most valuation methods require knowledge of a dose-response function relating varying levels of pollutants to physical damages. With this method, the first step in valuing environmental damages is to establish a functional relation between changes in residual flows and various measures of ambient environmental quality. Changes in environmental quality can then be related to the flow of environmental services to individuals and to economic welfare. For example, in the acid deposition case, an increase in acid precursor emissions could cause a reduction in the pH of water bodies and a loss of suitable fish habitat (Figure 1). The loss of fish habitat results in a decline or extinction of fish populations, a reduction in the quality of recreational angling, and a loss in economic welfare (Figure 2). An alternative method for valuing environmental damages directly measures them by observing the response of individuals to varying levels of environmental quality.

Although the underlying economic theory is well developed, there have been few empirical studies of water pollution damages [28–31]. Water quality (WQ) can affect individual economic welfare either directly or indirectly. If WQ were sold in markets like normal commodities, the value of alternative WQ levels could be determined readily. Under certain assumptions, changes in consumers' willingness to pay for WQ would be an appropriate measure of the changes in economic welfare resulting from changes in WQ. However, in other cases it might be more appropriate to value the welfare loss from pollution as the amount necessary to compensate existing users of a resource for the loss of their rights to its use. While there are several theoretical measures for approximating the damages associated with a decline in WQ, there is strong justification for the use of Marshallian consumer surplus as the measure of welfare loss because the income elasticity of

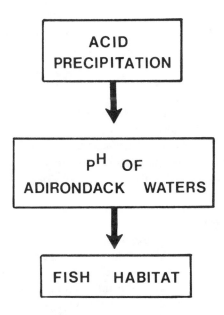

Figure 1. Environmental interaction function.

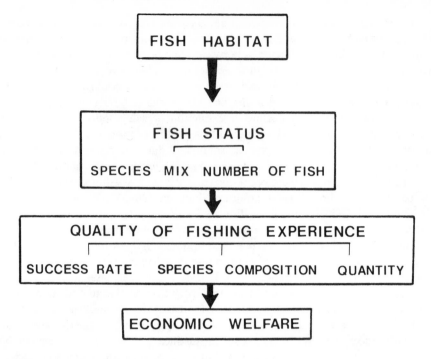

Figure 2.   Economic damage function.

demand for WQ is likely to be constant over most ranges of variation [28,32]. The consumer surplus measure—that is, the area under the demand curve for WQ—provides an approximate measure of economic welfare since it is bounded by the compensating and equivalent variation measures of potential welfare changes [28,29,31].

Since environmental services are not directly marketed, economic information concerning the demand for alternative levels of environmental quality must be derived. One method for estimating demand involves asking individuals directly or indirectly how much they would be willing to pay for an increase in the availability of environmental services (or, alternatively, an improvement in environmental quality). Another method relies on observed relations between the quantity of certain marketed goods or services demanded and the level of environmental quality. For example, in the case of a recreational fishing site, use of the site can be related to the price for its use and to certain angler and site characteristics, including water quality at the site. Assuming that use of the fishery is a necessary prerequisite for enjoyment of a given level of WQ, then the increase (decrease) in demand for days at the fishery as a result of an improvement (deterioration) in WQ will be a measure of the benefits (damages) of a change in WQ. A variant of this method is utilized in the current study.

In Figure 3, D_1 is the visit demand function for the site prior to a deterioration in WQ. At a price of $OP per day, OX_1 days will be demanded and the net

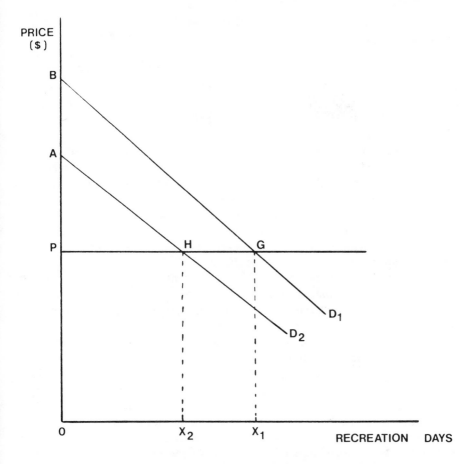

Figure 3. Demand for recreation and the damages from water pollution.

economic value of the site will be the area in GBP. With a deterioration in WQ, users' marginal willingness to pay would decrease to the demand curve D_2. At $OP the number of days would decrease to X_2 and the site's net economic value would be reduced to the area APH. The loss in economic welfare from increased pollution is the area between the demand curves that corresponds to the alternative levels of environmental quality (ABGH) [33]. This approach requires a method for estimating demand for the recreation site and also a way to determine the extent of the demand shift in response to a change in environmental quality. Ideally, an environmental quality variable could be included directly in the individual utility function to estimate how willingness to pay for a recreational site would vary with changes in the level of environmental quality. This approach is appealing from a theoretical standpoint, but its use is limited to cases where environmental quality can be measured accurately and individual responses to changes in environmental quality levels can be determined.

Unique Aspects of Valuing Acidification
Damages in the Adirondack Fishery

Several features make valuing damages different for acidification than for the typical water pollution case. The Adirondack fishery itself has unusual characteristics, and there are also problems in relating increased acidification to angling activity. In addition, the dose-response function appears to be nonmarginal, which makes the standard economic framework inappropriate.

The traditional economic approach to valuing environmental damages relies on incremental changes in environmental quality and the associated willingness to pay for this quality. However, pH as a measure of acidity is not likely to be directly perceived by anglers. Even if it were possible to observe directly pH (for example, as turbidity and odor can be observed), it is not clear how anglers' perception of pH would correlate with a scientific measure of pH [28,34]. This suggests that it would be inappropriate to include an environmental quality variable reflecting water pH directly in anglers' utility functions.

Another feature is the growing body of scientific evidence suggesting that damages related to acidification, particularly those in aquatic environments, are characterized by strong "all-or-nothing" elements. Henriksen [35] has argued that acidification of surface waters by atmospheric acid inputs can be viewed as a large-scale titration in which lakes can accept additional acid inputs for a period of time during which bicarbonates and other buffers are gradually depleted. Lakes then enter a transition phase characterized by severe pH fluctuations and significant damages to aquatic life. This hypothesis is supported by research in both Scandinavia [12,13] and North America [7].

Crocker and Forster [36] have argued that application of marginal analysis in the acid deposition case may not lead to optimal control strategies, since equating marginal benefits with marginal costs will fail to yield an economically efficient level of control if marginal damages do not increase monotonically with acidity. A nonconvex damage function implies that incremental changes in pH cannot be valued in terms of their marginal impact on willingness to pay since the flow of environmental services may only be affected if pH surpasses some critical level. Once this critical level is attained, the marginal damage of further pH reductions decreases or may even be negligible.

For these reasons, this study views acidification damages as causing the loss of certain water bodies in the Adirondack fishery, based on the assumption that acid loading affects ambient levels of pH and the ability of water to support aquatic life. This approach is not marginal in the standard sense of a well defined environmental interaction function relating marginal pollution inputs to changes in the quality of the fishery. The nonmarginal nature of the acidification process, along with the difficulties in relating angling demands to success rates, suggests that it would be appropriate to view acidification as causing a reduction in available angling acreage rather than in terms of incremental reductions of fish harvest rates. Furthermore, the nature of the Adirondack fishery is such that elimination of angling opportunities even in some of the remote high elevation water bodies may be of greater significance than marginal changes in success rates or fish harvest.

METHODS

Modeling Fisheries Demand in the Adirondacks

In principle, there are a number of possible approaches to estimating the value of pollution damages to a recreational fishery. However, due to the special characteristics of the acid deposition problem and the unique attributes of the Adirondack fishery, this study compares the economic value of the recreational fishery in the Adirondack region without substantial acidification damage to the fishery's value after angling opportunities are diminished because of acidification damage. An economic model was formulated to assess the current value of the Adirondack recreational fishery and to estimate the reduction in its value resulting from the loss of habitat due to acidification.

The travel cost method is widely used for estimating demand and net economic value for outdoor recreation resources [37]. This method infers willingness to pay for a recreation site by observing the frequency of visits from a number of points of origin located at different distances from the site. The first step of the travel cost procedure predicts visitation to a site as a function of travel costs and other factors. A demand curve is then derived showing how use would vary in response to a price charged for use of the site. Early applications of the travel cost procedure usually concerned a single site in isolation, assuming that there were no substitutes for the site. However, since this yields biased estimates if relevant substitution possibilities exist, a measure reflecting the availability of substitutes should be included when appropriate. The usual practice takes into account substitute sites within the same activity classification (intraactivity substitution), but substitution within a broader classification of recreational activities has also been considered. The diversity of angling opportunities in the Adirondack fishery, ranging from remote, high-elevation brook trout ponds to more accessible warmwater lakes, suggests that the value placed on a particular fishery site will depend on a substantial set of possible substitutes both within a particular type of fishery and across different types.

Both types of substitution effects could be accounted for in a model incorporating a system of demand equations where each equation represents an actual site and includes, as a regressor, a measure reflecting substitute sites [38,39]. This would result in a system of demand equations for interrelated recreation sites and would deal appropriately with the effects of substitutes on the demand for a particular site. However, observations of individual sites in the present study would involve a system of more than 3000 demand equations, entailing significant computational difficulties and most likely resulting in meaningless statistical relationships. A single equation model could also be used to predict the demand for a particular type of activity at any site on the basis of its own price and an index of substitute prices [40]. The substitute price index would include the price of trips to all competing sites whether or not they were in the same activity classification. However, treating substitutes in this manner would be equivalent to assuming that a substitute site in a different activity is a perfect substitute for an alternative site in the same activity.

The approach in this study represents a compromise between the single-equation technique and a complete system of demand equations. The model is formalized as:

$$D_i^1 = f\left(\sum^k \frac{A_k^1}{P_{ik}^1}, \sum^k \frac{A_k^2}{P_{ik}^2}, \ldots, \sum^k \frac{A_k^m}{P_{ik}^m}, SE_i\right) \tag{1}$$

$$\vdots$$

$$D_i^m = f\left(\sum^k \frac{A_k^m}{P_{ik}^m}, \sum^k \frac{A_k^1}{P_{ik}^1}, \ldots, \sum^k \frac{A_k^{m-1}}{P_{ik}^{m-1}}, SE_i\right) \quad \begin{array}{l} i = 1, \ldots, n \text{ origin zones} \\ j = 1, \ldots, m \text{ types of fisheries} \\ k = 1, \ldots, l \text{ sites} \end{array}$$

where D_i^j = demand (days per angler population) from zone i for fishery-type j

p_{ik}^j = price (time and money) of traveling from origin zone i to site k to engage in fishery-type j

A_k^j = measure of available water acreage at site k for fishery-type j (acreage for coldwater and other lakes, number of streams for streams)

SE_i = vector of socioeconomic and preference variables for the angler population in zone i

The value of a particular site cannot be inferred from this model, since the demand for a particular type of fishery is related to a composite site availability/price index for each type of fishery. However, this model accounts for both intraactivity substitution (across sites within the same fishery classification) and interactivity substitution (across fisheries, e.g., coldwater to warmwater) by measuring the change in total demand for a fishery-type in response to a change in the level of a number of indices, each reflecting available angling opportunities within a specific type of fishery.

Modeling Acidification Damages

Ideally, scientific information pertaining to estimates of habitat loss could be combined with information on the value of the habitat to anglers to develop an economic damage function relating losses in economic welfare to increased acidification of the fishery. This is impossible at present, given the lack of scientific data regarding acidification impacts as well as difficulties in modeling the full range of substitution possibilities that would be necessary to assess accurately the value of the habitat.

Earlier discussion has emphasized the importance of accounting for relevant substitution possibilities in estimating the economic value of the Adirondack fishery. In estimating the loss from acidification, however, sites within the Adirondacks that continue to support viable fish populations may or may not represent "relevant" substitution possibilities. For one thing, recent water chemistry and creel census data are available for only a limited sample of Adirondack waters, so acidification damage may in fact be more widespread within the fishery than current scientific evidence has shown. This suggests that intricate modeling of substitution possibilities may be inappropriate for valuing losses from acidification. Furthermore, regardless of scientific evidence, the angling public may perceive the acidification problem to be more widespread than it actually is and may alter their visitation to the fishery accordingly. This could result in a change in the entire character of visitation patterns within the fishery and possibly even cause diversion of angling activity away from the Adirondacks to other regional fisheries. The standard approach (within a partial equilbrium framework) assumes that significant substitution will occur within the Adirondack fishery as specific angling opportunities are lost, thus partially compensating for the loss of particular sites from acidification. It is also assumed that there would be no significant diversion of anglers to alternative fisheries outside the Adirondack region.

   To estimate the current economic losses to recreational anglers resulting from acidification of Adirondack waters, this study relies on an estimate of the fishery's value in a base year (in current dollars) and assumes that the waters found to be acidified since the base year are representative of recent habitat losses from acidification. The loss in economic welfare to anglers can be estimated by determining the reduction in the value of the Adirondack fishery to anglers after water acreage has been lost as a result of acidification. Since angling demands are related to available opportunities for each type of fishery via the water acreage variable in Equation 1, the loss in angling usage and value will be determined as substitution occurs to the remaining habitat subsequent to acidification damage. This approach is consistent with the scientific aspects of acidification, which suggest that marginal analysis based on changes in pH may be inappropriate due both to nonlinearities in the environmental interaction function and to the pervasiveness of the acidity impact within the regional fishery. However, the model is developed under the most optimistic set of assumptions because it assumes that significant substitution will take place within the Adirondack fishery as sites become acidified and also assumes that there will be no diversion of anglers to fisheries outside the Adirondack region.

EMPIRICAL IMPLEMENTATION

Participation Demand Equations

For empirical purposes, Equation 1 may be formulated as:

$$D_i^j = \alpha_j + \beta_{jj} \sum_{k=1}^{l} \frac{A_k^j}{P_{ik}^j} + \sum^{m \neq j} \beta_{jh} \left(\sum_{k=1}^{l} \frac{A_k^h}{P_{ik}^h} \right) + \theta_j \, SE_i + \epsilon_{ij} \qquad \begin{array}{l} (i = 1, \ldots, n) \\ j, h = 1, \ldots, m) \\ (k = 1, \ldots, l) \end{array} \qquad (2)$$

where α_j, β_{jj}, β_{jh} and θ_j are parameters to be estimated, and ϵ_{ij} is a zero mean error term. The β_{jj} parameter reflects the influence of available opportunities for a particular type of fishery on the per capita demand for that fishery. The hypothesized sign for this parameter is positive given the formulation of the opportunities index, which measures available water acreage at a vector of distance prices. Additional acreage or lower distance prices should result in increased demand. The β_{jh} parameters are intended to capture the influence of substitute angling opportunities on demand for a specific type of fishery and should exhibit a negative sign. It may be reasonable to expect that $\beta_{jh} = 0$ for some jh because differences in species composition, type of equipment required and the like may result in little or no substitution possibilities across certain types of fisheries.

Origins were used as units of observation in this study since information concerning individual income and travel costs was incomplete. Angling sites were classified as either ponded waters or streams and aggregated by 15-min quadrangles to form composite sites. Ponded waters were further subdivided into two classifications: "coldwater," referring to lakes and ponds managed primarily for salmonid species, and "other," which includes both warmwater and two-story lakes and ponds.

A demand equation was estimated for each of three fishery types relating days per angler population from each origin zone to the relevant availability/price indices as well as to socioeconomic variables. The demand equation for each of the ponded fishery types included availability/price indices for both types of ponded fisheries since they are likely to be substitutes. There is also likely to be some substitution between ponded waters in the "coldwater" classification and streams since the majority of the latter contain salmonid species as the primary resident game fish. Substitution between streams and "other" lakes and ponds, however, is much less likely to be important. In any event, lack of information pertaining to acreage or shoreline miles as a measure of availability of streams prevented their explicit consideration as substitutes for either classification of ponded waters. Moreover, it is questionable whether any such measures of stream availability would be the proper basis for comparing stream angling opportunities with ponded water acreage. As such, some bias in the estimated coefficients is possible to the extent that substitution takes place between streams and either classification of ponded waters.

The available opportunities index for the ponded fishery-types was defined as:

$$AP = \sum^k \frac{A_k^j}{P_{ik}} \tag{3}$$

where A_k^j = available acreage of fishery types j at site (quadrangle) k
$\quad\quad\quad$ P_{ik} = measure of travel costs, with time and distance as components, from origin i to site k

The index reflecting stream angling opportunities was defined as:

$$STAP = \sum^{k} \frac{N_k}{P_{ik}} \tag{4}$$

where N_k = number of streams at site (county) k
P_{ik} = measure of travel costs from county i to site k

The number of streams was used as an availability measure since other stream data were not available for this analysis.

The demand model presented above assumes that angling opportunities for alternative fishery types are likely to be substitutes for one another. Failure to consider relevant substitute opportunities will result in biased estimates of demand and economic value for a particular fishery type. For example, if geographically close substitutes were ignored, the measured demand relationship would exhibit a greater degree of inelasticity than the true relationship, since observed visitation rates will be relatively insensitive to higher distance prices. Substitution between different types of ponded waters in the Adirondack fishery is likely to be extensive and therefore was explicitly considered in the model. Substitution possibilities between streams and ponded waters were not explicitly considered due to data inadequacies and differences in the nature of these fisheries.

In addition to the relevant "price" variables, socioeconomic variables are usually considered to be demand determinants since they may reflect ability-to-pay or the strength of tastes and preferences. Of the numerous possible socioeconomic variables, income and species preference were used in this model. The mean annual income (INC) of respondents by origin zone was included as an explanatory variable in all three demand equations. A variable reflecting anglers' strength of preference for coldwater species (CWPREF), defined as the percentage of anglers from each origin zone whose "two most preferred species to fish for" were coldwater species, was included as a demand determinant for the coldwater fishery type. Preferences for the other ponded waters classification could not be easily measured since it included both warmwater and coldwater species. Preferences for stream fishing vs other types of angling were also unavailable. Thus, a preference variable was included for only one of the three classifications of fishery types.

Several data sources were used for this study. Angler income, preferences and travel costs were determined from a 1976 survey of licensed New York resident anglers [41]. Travel distances were based on measured mileage from the midpoint of each county of origin to the midpoint of each composite site (quadrangle or county). These measured distances were used to construct travel costs reflecting both the money and time price of traveling to a site. Data concerning the availability, classification, and characteristics of fishing sites were derived from Pfeiffer [23] and Pfeiffer and Festa [14].

Single-equation analyses of each demand relationship would generate unbiased and efficient parameter estimates if the demands for the three types of fisheries were related to the same set of variables. However, if unique determinants of demand exist or if the fisheries cannot be considered as viable substitutes, the disturbance terms across equations are likely to be correlated. With this system of interrelated demand, single-equation estimation procedures may lead to inefficient parameter estimates. A preliminary analysis of ordinary least-squares residuals indicated a strong degree of correlation among the error terms of the separate equations. Accordingly, Zellner's [42] seemingly unrelated regression (SUR) technique was used for estimation purposes, since it improves efficiency by accounting for contemporaneous correlations between the error terms of each equation. As noted above, some bias may exist if substitution between streams and ponded waters is important. In any event, the demand model was estimated both with and without the explicit consideration of substitution between ponded fishery types in order to illustrate the extent of the bias in the demand relationships and ultimate consumer surplus measures when relevant substitution possibilities are ignored.

The appropriate functional form for estimating recreation demand equations remains an unsettled issue. A linear format has been employed in applications relying on a system of demand equations, facilitating the imposition of restrictions that adhere to the symmetry conditions to obtain a unique measure of economic welfare. Both linear and logarithmic formulations have been used in single-equation applications. While reasonable justifications can be provided for any of the alternative functional forms, the double-log format has more intutitive appeal by virtue of its concern with percentage changes rather than absolute effects. Since this specification produced more significant and stable parameter estimates and greater explanatory power than the other forms, it was chosen as the basis for deriving the results. It should be noted, however, that consumer surplus measures may be highly sensitive to the choice of functional form [43].

RESULTS

Visitation and Demand

The results of the empirical model (Equation 2), estimated using the SUR technique, are presented in Table III. These results parallel the ordinary least-squares estimates (not shown here) except in the sense that they show a general improvement in efficiency as implied by higher t-statistics for the explanatory variables. The model was estimated both with and without explicit consideration of interfishery substitution to show the extent to which the coldwater (CW) and other ponded waters (LK) serve as substitutes for one another.

In general, the results confirm expectations, with visitation to each fishery significantly influenced by its own price, angler income and, to a lesser extent, strength of angler preference for species of fish (coldwater only). The results in Table III show that an increase (decrease) in angling opportunities for substitute types of angling will decrease (increase) the demand for both coldwater and other ponded angling opportunities. However, the coefficient of the substitute opportunities index was statistically insignificant in each case. This suggests that interfishery substitution in the Adirondack fishery is relatively unimportant, perhaps because of ample opportunities for all types of fishing across the entire Adirondack region.

Table III. SUR Results[a]

| | Fishery Type | | |
	Coldwater	Other Lakes/Ponds	Streams
With Interfishery Substitution			
Constant	−7.3664 (3.40)[b]	−21.4235 (4.51)	−7.9772 (2.64)
CWAP	+0.6174 (4.08)	−0.4245 (1.19)	
LKAP	−0.1407 (1.14)	+1.1282 (3.87)	
STAP			+0.6415 (7.55)
Income	+0.4766 (2.32)	+1.593 (3.50)	+0.7503 (2.43)
Without Interfishery Substitution			
Constant	−7.2896 (3.42)	−23.23 (5.06)	−8.20 (2.72)
CWAP	+0.4668 (7.80)		
LKAP		+0.8075 (7.53)	
STAP			+0.6568 (7.84)
Income	+0.4483 (2.23)	+1.7717 (4.04)	+0.7716 (2.51)
CWPREF	+0.5122 (1.39)		

[a] Dependent variable is days per angler population.
[b] Absolute values of t-statistics are in parentheses.

If substitution occurs in response to diminished angling opportunities, it is likely to take place within a particular fishery classification rather than across fishery types. These findings should not be taken as conclusive, however, in that multicollinearity between the indices for the different fisheries may be partially responsible for the insignificance of the substitute opportunities' parameter in each case. The results in Table III show that the measured elasticity of site visitation to variations in travel costs (price) is reduced if possible interfishery substitution between the different types of ponded waters is ignored. This result is consistent with theoretical expectations and suggests that estimates of the fishery's economic value to anglers would be exaggerated unless the economic model of the fishery reflects substitution properly (despite the insignificance of these variables). A further implication is that it would be desirable to determine somehow the extent to which substitution occurs between streams and ponded waters.

Net Economic Value of the Adirondack Fishery

The results in Table III were used to estimate the economic value of the Adirondack fishery. Each separate equation from the demand model (Equation 2) shows how angler visitation is related to variations in the index of acreage/travel costs to all sites within a particular fishery classification. These results were used to generate an aggregate demand curve for each fishery type by assuming that incremental

Table IV.   Adirondack Recreational Fishery Net Economic Values (1982 $)

Net Economic Value	Coldwater	Lakes	Streams	Entire Fishery
With Interfishery Substitution				
Total	2,680,955	22,044,855	14,104,970	38,319,595
Per angler-day	17.01	23.27	31.09	24.58
Without Interfishery Substitution				
Total	3,910,601	37,442,595	13,802,869	55,068,556
Per angler-day	24.80	39.52	30.43	35.33

fees are simultaneously imposed on all sites within the fishery classification until visitation is reduced to zero. The area under this aggregate demand curve was estimated to yield a measure of net economic value for each type of fishery. The economic value of the entire Adirondack fishery was then found by numerically aggregating across the different fishery classifications.

Table IV presents estimates of total economic value and economic value per angler-day for the Adirondack fishery. Two sets of estimates are presented to illustrate the differences in values that result from different assumptions concerning interfishery substitution in the demand models. As expected, the values for the ponded (CW and LK) fisheries, which are derived from the model that fails to consider interfishery substitution, are greater in magnitude because the aggregate demand curve is more steeply sloped than when substitutes are properly accounted for. The value of the stream fishery (ST) is affected only slightly with different formulations of the demand model because substitution between ponded waters and streams was not directly considered.

Losses from Acidification

The loss to anglers from acidification was approximated by calculating the reduction in economic value from reduced visitation to acidified ponded waters. For this purpose, it was necessary to compile two sets of availability/price indices. The first set, which was used to estimate the value of the fishery prior to acidification, was composed of pre-1976 data and included all waters with pH ≥ 5.0 that could be classified according to type of fishery. Another set of indices was compiled to reflect reduced angling opportunities as a result of acidification and therefore excluded ponded water bodies which exhibited a pH of <5.0 in post-1976 surveys of Adirondack waters [15]. These indices were then combined with the parameter estimates from the model of the Adirondack fishery to determine the loss in angling days and reduction in economic values as angling opportunities were diminished by acidification.

The resulting estimates of acidification-related losses in both angler visitation and net economic value are presented in Table V. Two sets of estimates are presented to illustrate how different assumptions concerning interfishery substitution possibilities affect estimated acidification losses. The results show losses of approximately $1.7 million per year in angler economic value if substitution between ponded fishery types is considered in estimating the fishery's economic value. It can be seen that greater losses in angler days are estimated when interfishery substitution is considered than when it is ignored. The larger loss of angler days in this case results from the greater measured degree of elasticity and the inconclusiveness regarding the statistical significance of the substitution parameters in both demand equations. However, because per-angling-day economic values are less with substitution considered, the estimated loss in economic value is less than if substitution possibilities are ignored.

There are several reasons why the estimates in Table V may understate actual economic losses to anglers in the Adirondack fishery. First, they do not include diminished stream angling opportunities [27]. Second, they are based only on specific waters found to be acidified in post-1976 surveys. The extent of acidification damage in the Adirondack fishery may be significantly greater than is indicated by current scientific information. It is possible, too, that some of the waters rencently found to exhibit a pH in the range of 5.0–6.0 may eventually exhibit a pH <5.0 because they are likely to have limited buffering capacity. Since waters in this pH range comprise approximately 30% of total acreage in the Adirondack fishery, the potential losses could be substantially higher than those in Table V.

It would be inappropriate to deal with additional habitat losses of this magnitude in a partial equilibrium framework. However, to provide information about possible economic losses if acidification damage were greater than currently known, another estimate of losses was performed assuming a doubling of known acreage losses within the composite sites which currently have waters in the critical state. The results are presented in Table VI. It can be seen that losses in both angler days and economic values increase substantially from the results in Table V (whether or not interfishery substitution is directly considered), but by different proportions for the two types of fisheries. The indicated loss in coldwater angling days implies that there would be more substitution within the coldwater fishery as habitat is

Table V. Annual Losses to Anglers from Acidification of the Adirondack Fishery

	Coldwater	Lake	Total
With Interfishery Substitution			
Loss in angler days	82,257	11,268	93,525
Loss in net economic value (1982 $)	1,398,797	262,157	1,660,954
Without Interfishery Substitution			
Loss in angler days	66,064	8,233	74,297
Loss in net economic value (1982 $)	1,638,705	325,335	1,964,040

Table VI.　Estimated Annual Losses to Anglers Assuming Loss of Twice Critical Average

	Coldwater	Lake	Total
With Interfishery substitution			
Loss in angler days	123,396	22,040	145,436
Loss in net economic value (1982 $)	2,098,374	512,774	2,611,148
Without Interfishery Substitution			
Loss in angler days	103,390	16,130	119,520
Loss in net economic value (1982 $)	2,564,568	637,393	3,201,961

reduced than within the other ponded fishery classification. The findings in Tables V and VI also suggest that, whereas the initial habitat loss affects visitation to the entire fishery, substitution within each fishery classification tends to increase as acidification damage becomes more widespread.

The preliminary nature of the data relating to stream acidification and the greater degree of uncertainty regarding the effects of pH fluctuations in streams on aquatic ecosystems prevent estimating the economic losses resulting from their acidification. Until a more precise relationship between acidification of streams and ponded waters is developed, estimating the value of acidification damages to streams would be premature. Even then, the conceptual problems involved in comparing stream and ponded water fishing opportunities so as to appropriately model the full range of substitution possibilities would have to be addressed before reliable estimates could be made.

The results of Tables V and VI yield a range of estimates of losses in economic welfare to anglers resulting from acidification damage. The estimates are all conservative, however, since they assume that there is no diversion of anglers away from the Adirondack fishery and that substitution within specific fishery classifications occurs as anglers become aware of acidification damage to particular sites. Given these assumptions, the annual losses to licensed New York resident anglers range from $1.7–3.2 million, depending on assumptions pertaining to habitat loss and substitution by anglers among fisheries. This estimate of damages is apart from the regional economic impact of reduced angler expenditures in the range of $1–2 million annually [44].

CONCLUSIONS

This chapter presents an estimate of economic losses to anglers in the Adirondack recreational fishery resulting from acidification damage. These results should be treated as lower-bound estimates for several empirical as well as conceptual reasons. First, the estimates of the Adirondack fishery's value are based only on licensed New York resident anglers fishing in waters open to public fishing. Second, they are based on the assumption that information concerning the effect of acidification on alternative fishing sites is known and accurately perceived by the angling population. Accordingly, anglers are assumed to alter their visitation patterns in response to diminished angling opportunities as reflected in the underlying economic model of the fishery. If anglers' perception of the extent of acidification is inaccurate, or if they lack information concerning specific sites that have been lost due to acidification, the loss from acidification predicted by this model will be incorrect. For example, if anglers perceive that the damage from acidification is fairly widespread, then a much more drastic change in visitation patterns will result than the empirical model predicts. Third, it is possible that acidification damage is irreversible in its effects, so that today's actions effectively constrain future activities. Estimates of currently perceived damages that take into account only the reduction in economic welfare from the current generation's perspective may substantially understate actual damages. Finally, the results of this study pertain only to the currently observable effects of acidification on a limited portion of the Adirondack recreational fishery. Other, more subtle, and perhaps more widespread, effects of acidification on the aquatic ecosystem are either unknown or are not known with certainty. At the present time, there is no generally accepted procedure for dealing adequately with these situations within the framework of cost-benefit analysis.

ACKNOWLEDGMENTS

This research has been supported in part by the Office of Water Research and Technology, U.S. Department of Interior (B-103-NY) and by the New York Sea Grant Institute. Data were provided by the Department of Environmental Conservation, State of New York. Donald Wilton provided excellent research assistance.

REFERENCES

1. Nriagu, J.O., Ed. *Sulfur in the Environment, Part I: The Atmospheric Cycle* (New York: John Wiley & Sons, Inc., 1978).
2. Nriagu, J.O., Ed. *Sulfur in the Environment, Part II: Ecological Impacts* (New York: John Wiley & Sons, Inc., 1978).
3. Galloway, J.N., G.E. Likens and E.S. Edgerton. "Acid Precipitation in the Northeastern U.S.: pH and Acidity," *Science* 194(12):722–723 (1976).
4. Oden, S. "The Acidity Problem—An Outline of Concepts," in *Proceedings of the First International Symposium on Acid Precipitation and the Forest Ecosystem*, L.S. Dochinger and T.A. Seliga, Eds. Technical Report NE-23 (Upper Darby, PA: USDA Forest Service, 1976), pp. 1–26.
5. Toribara, T.Y., N.W. Miller and P.E. Morrow, Eds. *Polluted Rain* (New York: Plenum Press, 1980).
6. "Acid Rain Information Book: Final Report," DOE/EP-0018, Office of Environmental Assessments, U.S. Department of Energy, Washington, DC (1981).
7. "Memorandum of Intent on Transboundary Air Pollution, Interim Report," Canada/United States Impact Assessment Work Group, Washington, DC (1981).

8. Dochinger, L.S., and T.A. Seliga, Eds. *Proceedings of the First International Symposium on Acid Precipitation and the Forest Ecosystem,* Technical Report NE-23 (Upper Darby, PA: USDA Forest Service, 1976).

9. "Ecological Effects of Acid Precipitation," EPRI Report EA-79-6-LD Electric Power Research Institute, Palo Alto, CA (1979).

10. Shriner, D.S., C.R. Richmond and S.E. Lindberg, Eds. *Atmospheric Sulfur Deposition: Environmental Impact and Health Effects* (Ann Arbor, MI: Ann Arbor Science Publishers, Inc., 1980).

11. Fromm, P.O. "A Review of Some Physiological and Toxicological Responses of Freshwater Fish to Acid Stress," *Environ. Biol. Fishes* 5(1):79–93 (1980).

12. Haines, T.A. "Acidic Precipitation and Its Consequences for Aquatic Ecosystems: A Review," *Trans. Am. Fish. Soc.* 110(6):669–707 (1981).

13. Almer, B., W. Dickson, C. Ekstrom, E. Hornstrom and U. Miller. "Effects of Acidification on Swedish Lakes," *Ambio* 3:30–36 (1974).

14. Pfeiffer, M.H., and P. Festa. "Acidity Status of Lakes in the Adirondack Region of New York in Relation to Fish Resources," FW-P168, New York State Department of Environmental Conservation, Albany, NY (1980).

15. Schofield, C.L. "Acid Precipitation: Effects on Fish," *Ambio* 5:228–230 (1976).

16. Beamish, R.J. "Acidification of Lakes in Canada by Acid Precipitation and the Resulting Effects on Fishes," *Water, Air, Soil Poll.* 6:501–514 (1976).

17. Beamish, R.J., and H.H. Harvey. "Acidification of the LaCloche Mountain Lakes, Ontario, and the Resulting Fish Mortalities," *J. Fish. Res. Board Can.* 29(8):1131–1143 (1972).

18. Gorham, E. "Acid Precipitation and Its Influence on Aquatic Ecosystem—An Overview" in *Proceedings of the First International Symposium on Acid Precipitation and the Forest Ecosystem,* L.S. Dochinger and T.A. Seliga, Eds., Technical Report NE-23 (Upper-Darby, PA: USDA Forest Service, 1976), pp. 425–458.

19. Hendrey, G.R., J.N. Galloway, S.A. Norton, C.L. Schofield, P.W. Shaffer and D.A. Burns. "Geological and Hydrochemical Sensitivity of the Eastern United States to Acid Precipitation," EPA-600/3-80-024, U.S. EPA, Corvallis, OR (1980).

20. Wright, R.F., N. Conroy, W.T. Dickson, R. Harriman, A. Henriksen and C.L. Schofield. "Acidified Lake Districts of the World: A Comparison of Water Chemistry of Lakes in Southern Norway, Southern Sweden, Southwestern Scotland, the Adirondack Mountains of New York and Southeastern Ontario," in *Proceedings of the International Conference on the Ecological Impact of Acid Precipitation* D. Drablos and A. Tollan, Eds. (Oslo, Norway: SNSF Project, 1980), pp. 377–379.

21. Driscoll, C.T., J.P. Baker, J.L. Bisogni and C.L. Schofield. "Effect of Aluminum Speciation on Fish in Dilute Acidified Waters," *Nature* 284:161–164 (1980).

22. Schofield, C.L., and R.R. Trojnar. "Aluminum Toxicity to Fish in Acidified Waters," in *Polluted Rain* T. Toribara, M.W. Miller and P.E. Morrow, Eds. (New York: Plenum Press, 1980), pp. 341–366.

23. Pfeiffer, M.H. "A Comprehensive Plan for Fish Resource Management Within the Adirondack Zone," FW-P142 New York State Department of Environmental Conservation, Albany, NY (1979).

24. Blake, L.M. "Liming Acid Ponds in New York," *N.Y. Fish Game J.* 28(2):208–214 (1981).

25. Robinson, G.D., W.A. Dunson, J.E. Wright and G.E. Mamolito. "Differences in Low pH Tolerance Among Strains of Brook Trout (*Salvelinus fontinalis*)," *J. Fish Biol.* 8:5–17 (1976).

26. Pfeiffer, M.H. Personal communication (June 29, 1981).

27. Colquhon, J.R., J. Symula, M.H. Pfeiffer and J. Feurer. "Preliminary Report of Stream Sampling for Acidification Studies—1980," FW-P182 New York State Department of Environmental Conservation, Albany, NY (1981).

28. Freeman, A.M. *The Benefits of Environmental Improvement: Theory and Practice,* (Baltimore, MD: The Johns Hopkins University Press, 1979).

29. Feenberg, D., and E.S. Mills. *Measuring the Benefits of Water Pollution Abatement* (New York: Academic Press, Inc., 1980).

30. Freeman, A.M. "The Benefits of Air and Water Pollution Control: A Review and Synthesis of Recent Estimates," unpublished report prepared for the Council on Environmental Quality, Washington, DC (1979).

31. Maler, K.G., and R.E. Wyzga. "Economic Measurement of Environmental Damage," Organization for Economic Cooperation and Development, Paris (1976).

32. Willig, R.D. "Consumers' Surplus Without Apology," *Am. Econ. Rev.* 66(4):589–597 (1976).

33. Maler, K.G. *Environmental Economics: A Theoretical Inquiry* (Baltimore, MD: The Johns Hopkins University Press, 1974).

34. David, E.L. "Public Perceptions of Water Quality," *Water Resources Res.* 7(3):453–457 (1971).

35. Henriksen, A. "A Simple Approach for Identifying and Measuring Acidification of Freshwater," *Nature* 278:542–545 (1979).

36. Crocker, T.D., and B.A. Forster. "Decision Problems in the Control of Acid Precipitation: Nonconvexities and Irreversibilities," *J. Air Poll. Control Assoc.* 31(1):31–37 (1981).

37. Clawson, M., and J.L. Knetsch. *Economics of Outdoor Recreation* (Baltimore, MD: The Johns Hopkins University Press, 1971).

38. Burt, O.R., and D. Brewer. "Estimation of Net Social Benefits from Outdoor Recreation," *Econometrica* 39:813–827 (1971).

39. Cicchetti, C.J., A.C. Fisher and V.K. Smith. "An Econometric Evaluation of a Generalized Consumer Surplus Measure: The Mineral King Controversy," *Econometrica* 44:1259–1276 (1976).

40. Cesario, F.J., and J.L. Knetsch. "A Recreation Site Demand and Benefit Estimation Model," *Regional Studies* 10:97–104 (1976).

41. "1976–77 New York Resident Angler Survey," Bureau of Fisheries, New York State Department of Environmental Conservation, Albany, NY (1977).

42. Zellner, A. "An Efficient Method of Estimating Seemingly Unrelated Regressions and Tests for Aggregation Bias," *J. Am. Stat. Assoc.* 57:348–368 (1962).

43. Ziemer, R.F., W.N. Musser and R.C. Hill. "Recreation Demand Equations: Functional Form and Consumer Surplus," *Am. J. Agric. Econ.* 62:136–141 (1980).

44. Menz, F.C. "A Preliminary Estimate of the Economic Significance of Acid Deposition to the Adirondack Recreational Fishery," report prepared for the New York State Department of Law, Clarkson College, Potsdam, NY (1981).

CHAPTER 10

Transferable Discharge Permits and Profit-Maximizing Behavior

John T. Tschirhart

Acid deposition represents one of the most recent and vexing environmental problems. Long-range transport of air pollutants dramatically increases the difficulty of adopting measures to clean the air. Pollutants that cross state and national boundaries obscure their own sources, and government authorities know neither how to deal with the problem nor even whom to deal with. Moreover, intergovernmental coordination of efforts to clean the air has not been successful in the past, and future prospects appear no better. But solutions clearly require cooperation given that Maine receives Ohio's pollutants, Canada receives deposition from the United States, and Scandanavia confronts pollutants from England.

The purpose of this chapter is to explore one possible remedy for acid deposition specifically, and for pollution in general. The remedy is to use transferable discharge permits (TDP) that place a ceiling on total pollution emissions, and achieve that ceiling at minimum total cost to the sources. Crocker [1] discussed TDP as a remedy as far back as 1966, followed by Dales [2] in 1968. However, using TDP to combat acid deposition requires more knowledge than is currently available. Roughly, a certain number of permits may be allowed for a particular airshed, and at the local level this could be handled relatively easily. However, if the airshed extends from the Ohio Valley to New England, many sources and government agencies need to be involved. Nevertheless, the assumption here is that TDP are to be used. Given this, the chapter addresses potential impediments to a successful program that are inherent in government and private institutions.

To control pollution, economists have traditionally considered two distinct methodologies: effluent charges and effluent standards. With effluent charges, a regulatory agency levies a charge per unit of emissions on the polluting firms. Effluent standards work by giving firms emission quotas that they cannot exceed. Economists have favored the former approach on the basis of efficiency. With the proper charges, the optimum amount of pollution is obtained at minimum cleanup cost. This is because all firms equate the value of their marginal product of emissions to the charge, which, in turn, is equal to the marginal damage to society of the emissions. With standards, the optimum level is essentially mandated, but there is no guarantee that it will be achieved at minimum cost. Achieving the optimum with either system requires the agency to have an unrealistically large amount of data about damage functions. For this reason, and because effluent charges have never gained acceptance among lawyers and politicians, this system has not been attempted for air pollution in the United States.

Thus, the standards approach remains; in the absence of damage functions, not only is the optimum level of pollution unlikely, but the level attained is not attained at least cost. Transferable discharge permits offer a partial remedy. The agency distributes permits or entitlements to polluters that allow them to emit at determined levels. By controlling the number of entitlements issued, total emissions are determined. The firms can buy and sell the permits to one another, and it is this market for permits that drives the cost of attaining the total emission level to a minimum. If a firm wants to increase production and thereby emissions beyond what is allowed by its entitlements, it has a choice between cleanup costs or buying additional entitlements. However, it will only be able to buy entitlements from another firm whose cleanup costs must be less than the price at which the entitlements are sold. Therefore, the firm that does the cleaning is the one with lower cleanup costs.

This is a simple and short explanation of TDP; more detail can be found in recent papers by Tietenberg [3] and Atkinson and Tietenberg [4]. An actual system for TDP being implemented by the U.S. Environmental Protection Agency (EPA) is very similar to that described above [5]. The system is called banking, since firms that invest in pollution control and lower their emissions beyond their quota earn emission credits that can be either banked and saved for the future or sold to other firms. A model of a polluting firm is presented in the next section that captures the essential elements of the EPA system.

The merits of a TDP system are rigorously described by Montgomery [6]. He sets up a model of independent, profit-maximizing, polluting firms in a region that is small relative to the entire economy. The smallness ensures that changes in outputs in the region due to emissions standards will have a negligible impact on the economy and, therefore, output and input prices in the region are unaffected. Because the firms are profit maximizers, they are also cost minimizers for every output level produced. Since meeting the emissions standards is one form of cost, these costs are also minimized, and a regionwide, minimum-cost solution to pollution control is established.

An important assumption in Montogomery's analysis, and one that forms the subject of this chapter, is that firms attempt to and are able to maximize profits. There are two general reasons why this assumption may be violated: (1) firms may be constrained in their attempt to maximize profit and (2) profit maximization may not be the firm's goal or, even if it is the goal, uncertainty and organizational complexity may make it unattainable. The second reason is taken up below. The first reason arises in regulated industries and principally in the electric utility industry. Public utilities have their profits regulated in state public service commissions in the United States, and they cannot be treated in the same way as unregulated profit-maximizing firms. This tends to thwart a minimum cost solution to pollution control as is demonstrated below.

Before exploring the reasons why TDP may not always work as desired, a model of a profit-maximizing firm is presented. This model demonstrates how the system works under ideal conditions.

PROFIT MAXIMIZATION AND PARTICIPATION IN TDP MARKETS

A model of a profit maximizing firm is presented in this section. The firm combines capital and labor inputs to produce a single output, and pollution is a by-product of the manufacturing process. The firm is only allowed a level of pollution emissions as prescribed by a pollution control agency. Formally, the firm is granted an initial endowment of entitlements that place a ceiling on emissions. The ceiling level is less than what the firm would emit in the absence of regulation; therefore, the control agency creates a binding constraint on the firm's activities. To meet this constraint, the firm can either purchase additional entitlements to emit, or adopt a technology that will decrease emissions (e.g., scrubbers). Another option for the firm is to sell part or all of its initial endowment of entitlements; but this would require an increased level of antipollution technology.

There is a market for the purchase and sale of entitlements among firms. Firms wishing to buy or sell negotiate with firms wishing to sell or buy. If there are numerous firms wishing to buy or sell at any given time, and information on prices is readily available, the market for entitlements will be competitive and a uniform price per unit of entitlement will prevail. If price varies across selling firms there would be incentive for arbitrage, which in turn would cause prices to seek a uniform level. However, if the number of firms buying or selling is few in number, or information about prices is less complete, price dispersion may exist. The assumption used in the model is that firms do enter into negotiations when buying or selling, and obtaining information about prices is not costless. Therefore, if c is a price in the market, a buying firm actually pays c^+ per unit purchased where $c^+ > c$. The difference, $c^+ - c$, represents the transaction costs of negotiating and seeking information about prices. Similarly, the same firm selling would receive $c^- < c$ per unit sold, because of the costs incurred in seeking out higher bidders. This says simply that for any firm, the revenue generated from selling an entitlement is less than the cost of purchasing an entitlement due to transaction costs.

Profit for the firm is revenue minus cost where revenue may include income generated from sales of entitlements, and costs may include the purchase of entitlements. Profit is given by

$$\pi = P(y)y - rK - wL - rK^s - c^+E^p + c^-E^s \qquad (1)$$

where   y = output of the firm

       P(y) = price of the output which is a function of output

K = capital used in the production process
r = price of capital
L = labor used in the production process
w = price of labor
K^s = capital used to reduce pollution which will be referred to as scrubbers
E^P = number of entitlements purchased
E^s = number of entitlements sold
c^+ = cost of purchasing an entitlement
c^- = revenue from selling an entitlement

Output is a function of the capital and labor used in the production process; therefore, y can be replaced by

$$y = f(K,L) \tag{2}$$

Marginal products of capital and labor, given by f_K and f_L, are assumed to be positive. Note that pollution control capital is separable from capital used in the production of output. This assumption may appear too strong in cases where pollution control activity changes the production technology itself. However, the assumption simplifies the presentation, and including K^s as an argument of f does not alter the principal results derived below. Also note that the firm has control over price, which implies some monopoly power. For a firm in a competitive output market, $P(y)$ is replaced by a parameter P. The results that follow pertain to either case. The derivative of $P(y)$, written as P', is assumed to be negative.

The firm is granted E° as an initial endowment of entitlements by the control agency. How this endowment is determined will not be covered here. Total emissions are nonnegative and will depend upon output and the scrubbers used. Thus,

$$g(K^s,y) \geqq 0 \tag{3}$$

is total emissions or a pollution function for given values of K^s and y. The partial of Equation 3 with respect to K^s, denoted g_s, is negative, indicating that scrubbers reduce pollution; and the partial of Equation 3 with respect to output Y, g_f, is positive since pollution increases with output. The control agency's imposed constraint is then

$$g(K^s,y) \leqq E^\circ + E^p - E^s \tag{4}$$

or total pollution cannot exceed the initial endowment of entitlements plus any purchased entitlements minus any sold entitlements. Finally, for feasibility the number of entitlements sold cannot exceed the sum of the initial endowment plus purchases,

$$E^s \leqq E^\circ + E^p \tag{5}$$

but this is already implied by Equations 3 and 4.

The problem confronting the firm is to select values of K, L, K^s, E^p and E^s that maximize π subject to Equations 2, 3 and 4. Substituting Equation 2 into Equations 1, 3 and 4, and using δ and γ as Lagrange multipliers for constraints Equations 3 and 4, respectively, the Kuhn–Tucker conditions for a maximum are as follows:

$$K: f_K[P + P'f] - r + \delta g_f f_K - \gamma g_f f_K \leqq 0 \tag{6}$$

$$K[f_K[P + P'f] - r + \delta g_f f_K - \gamma g_f f_K] = 0 \tag{7}$$

$$L: f_L[P + P'f] - w + \delta g_f f_L - \gamma g_f f_L \leqq 0 \tag{8}$$

$$L[f_L[P + P'f] - w + \delta g_f f_L - \gamma g_f f_L] = 0 \tag{9}$$

$$K^s: -r + \delta g_s - \gamma g_s \leqq 0 \tag{10}$$

$$K^s[-r + \delta g_s - \gamma g_s] = 0 \tag{11}$$

$$E^p: -c^+ + \gamma \leqq 0 \tag{12}$$

$$E^p[-c^+ + \gamma] = 0 \tag{13}$$

$$E^s: c^- - \gamma \leqq 0 \tag{14}$$

$$E^s[c^- - \gamma] = 0 \tag{15}$$

$$\delta: g(K^s, f(K,L)) \geqq 0 \tag{16}$$

$$\delta[g(K^s, f(K,L))] = 0 \tag{17}$$

$$\gamma: E^\circ + E^p - E^s - g(K^s, f(K,L)) \geqq 0 \tag{18}$$

$$\gamma[E^\circ + E^p - E^s - g(K^s, f(K,L))] = 0 \tag{19}$$

$$K, L, K^s, E^p, E^s, \gamma, \delta \geqq 0 \tag{20}$$

To analyze these conditions, emissions are assumed to be positive. This is realistic in that the firm cannot reduce emissions to zero without shutting down production. Therefore, $g(K^s, y) > 0$ and from Equations 16 and 17, $\delta = 0$.

Equations 6 through 20 admit the following interpretations. First, regarding production, if positive output is produced and if both capital and labor are needed to produce, then K, L > 0 and both Equations 6 and 8 are satisfied by equalities. Combining Equations 6 and 8 yields

$$\frac{f_K}{f_L} = \frac{r}{w} \tag{21}$$

which is the usual condition for production efficiency.

Equation 6 can be written as

$$f_K[P + P'f] = r + \gamma g_f f_K \tag{22}$$

where the left side is the marginal revenue product of capital. The envelope theorem can be used to interpret the multiplier γ as the marginal profit of allowing an additional unit of emissions; therefore, the second term on the right side is the marginal profit from a change in emissions due to a change in output as capital is changed. This follows since g_f is the change in emissions from a change in output. Hence, the firm employs capital to the point where the marginal revenue from the last unit employed equals the sum of the marginal cost of that unit and the marginal profit loss from increased emissions. Figure 1 depicts this capital choice. The marginal revenue product is downward sloping, which is a necessary condition for profit maximization. Without an emission standard or with a standard that is nonbinding on the firm's actions so that $\gamma = 0$, the firm selects K' units of capital. However, K'' units are chosen with a binding emissions standard. That an emissions standard decreases the amount of capital chosen is not surprising, since it simply increases the cost of producing output. That is, increasing output via increased capital implies more emissions and more cost in controlling emissions. A similar story can be told for labor with Equation 8; thus, the firms's output unambiguously decreases with the introduction of an emission standard.

The method used by the firm to satisfy the emissions constraint is now examined. The firm satisfies the emission standard by assigning nonnegative values to one or more of K^s, E^p, and E^s. There are eight possibilities:

	1	2	3	4	5	6	7	8
K^s	+	+	+	+	0	0	0	0
E_p	+	+	0	0	+	+	0	0
E_s	+	0	+	0	+	0	+	0

At the outset of this section, the ceiling level of emission set by the control agency was assumed to be so low that the firm must take measures to meet the standard. In other words,

$$E^\circ - g(0, f(K,L)) < 0 \tag{23}$$

at the optimum values of K and L and with $E^p = 0$. This contradicts Equation 18 and a binding constraint on emissions eliminates possibility 8 and *a fortiori* possibility 7. The firm, at a minimum, must either purchase some additional entitlements or install scrubbers.

Next, note that $E^p > 0$ implies $\gamma = c^+$ from Equation 12, and $E^s > 0$

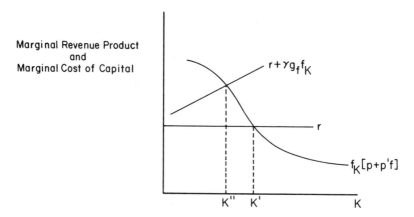

Figure 1. Optimum choice of capital.

implies $\gamma = c^-$ from Equation 14. Since this is impossible, the firm does not both buy and sell entitlements. This is obvious because $c^+ > c^-$ and the firm would lose on every entitlement purchased. This eliminates possibilities 1 and 5.

The remaining possibilities, 2, 3, 4 and 6, cannot be eliminated. Consider 2. Here $K^s > 0$ and $E^p > 0$ which implies from Equations 10 through 13 (recall $\delta = 0$) that

$$c^+ = -\frac{r}{g_s} \tag{24}$$

The right side is the marginal cost of scrubbers per unit of reduced emissions. Equation 24 requires that at the optimum the firm purchases entitlements and installs scrubbers until their marginal costs are equated. Figure 2 illustrates this

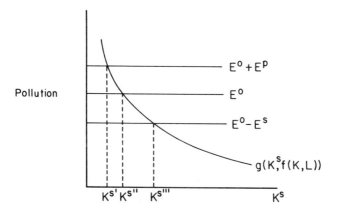

Figure 2. Optimum choice of scrubbers.

possibility. Emissions are drawn as a decreasing function of the scrubbers K^s. But emissions must be below a particular level; viz., the firm can emit E^o. This would require installing $K^{s''}$ of scrubbers. Rather than do this, the firm opts to purchase E^p of additional entitlements and only install $K^{s'}$ of scrubbers.

For possibility 3, $K^s > 0$ and $E^s > 0$ which implies from Equations 10, 11, 14 and 15 that

$$c^- = -\frac{r}{g_s} \qquad (25)$$

At the optimum the firm sells entitlements until the marginal revenue from selling is equal to the marginal cost of scrubbers. Figure 2 also illustrates this. The firm could employ $K^{s''}$ of scrubbers and satisfy the emission standard. Instead, the firm chooses to sell entitlements and purchase $K^{s'''}$ of scrubbers. The extra revenue obtained from the sale of entitlements more than compensates for $K^{s''} - K^{s'''}$ of scrubbers.

For possibility 4, $K^s > 0$. If Equations 12 and 14 are satisfied by inequalities, the result is

$$c^- < -\frac{r}{g_s} < c^+ \qquad (26)$$

Basically, entitlements are too expensive to purchase but too inexpensive to sell. Scrubbers are purchased to the point where the standard is met by $K^{s''}$ in Figure 2.

Finally, possibility 6 has $E^p > 0$. With Equation 10 satisfied by an inequality, the result is

$$c^+ < \frac{r}{g_s} \qquad (27)$$

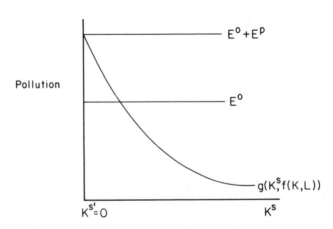

Figure 3. The optimum level of scrubbers is zero.

At the optimum, purchasing entitlements is less than the marginal cost of scrubbers, and the standard is met with the initial endowment plus purchased entitlements only—no scrubbers are installed. Figure 3 shows this result.

The important point in all these possibilities is that the firm is minimizing the cost of meeting the emission standard by responding to the marginal costs or revenues of the various alternatives. By equating the appropriate marginal signals, the optimum mix of scrubbers and entitlements is obtained. And when this firm is selling entitlements to another firm, the other firm's purchase is based on a cost-minimizing decision. Consequently, the overall emissions level set by the standards is achieved at minimum cost.

PUBLIC UTILITIES AND REGULATED PROFITS IN TDP MARKETS

Electric public utilities are a significant source of pollution and might be expected to participate in a market system for entitlements. However, the model in the previous section does not apply to utilities, because these firms are regulated by state public service commissions. The objective of profit maximization is the driving force behind a cost minimization solution to the emissions problem. To maximize profits, a firm must minimize costs of production for the output produced. But when a firm's rate of return is regulated, distortions of marginal costs can prevent the attainment of a least cost solution.

To illustrate this, a rate-of-return constraint is appended to the profit maximization model. The constraint requires that the utility's return to capital cannot exceed a fixed amount, and it is representative of how commissions regulate utilities. Averch and Johnson [7] were the first to employ this constraint in a seminal paper on utilities. Their results, later refined by Takayama [8] and discussed by numerous other authors, showed that when the constraint is binding, the utility employs more capital than cost minimization would require. This overcapitalization bias has been tested by several authors who have found some support for the theory [9–11].

In the present context, the constraint is as follows:

$$\frac{P(y)y - wL - c^+E^p + c^-E^s}{K + K^s} \leqq s \tag{28}$$

where s is the allowed rate of return and the left side is the return to capital. The scrubbers are assumed to be part of the rate base. This is true in all states; however, the timing as to when scrubbers are included may vary among states [12]. Some state commissions allow K^s in the rate base during construction, while others allow it in the rate base only after construction is completed. Also, s is greater than the cost of capital r, to allow some positive profit, but it is less than the return the firm would earn if it was unregulated. The Kuhn–Tucker conditions for this model are given by Equations 29 through 45, where λ is the Lagrange multiplier for the rate of return constraint.

$$K: f_K[P + P'f][1 - \lambda] - r + \delta g_f f_K - \gamma g_f f_K + \lambda s \leqq 0 \tag{29}$$

$$K[f_K[P + P'f][1 - \lambda] - r + \delta g_f f_K - \gamma g_f f_K + \lambda s] = 0 \tag{30}$$

$$L: [f_L[P + P'f] - w][1 - \lambda] + \delta g_f f_L - \gamma g_f f_L \leqq 0 \tag{31}$$

$$L[[f_L[P + P'f] - w][1 - \lambda] + \delta g_f f_L - \gamma g_f f_L] = 0 \tag{32}$$

$$K^s: -r + \delta g_s - \gamma g_s + \lambda s \leqq 0 \tag{33}$$

$$K^s[-r + \delta g_s - \gamma g_s + \lambda s] = 0 \tag{34}$$

$$E^p: -c^+ + \gamma + \lambda c^+ \leqq 0 \tag{35}$$

$$E^p[-c^+ + \gamma + \lambda c^+] = 0 \tag{36}$$

$$E^s: c^- - \gamma - \lambda c^- \leqq 0 \tag{37}$$

$$E^s[c^- - \gamma - \lambda c^-] = 0 \tag{38}$$

$$\delta: g(K^s, f(K,L)) \geqq 0 \tag{39}$$

$$\delta[g(K^s, f(K,L))] = 0 \tag{40}$$

$$\gamma: E^\circ + E^p - E^s - g(K^s, f(K,L)) \geqq 0 \tag{41}$$

$$\gamma[E^\circ + E^p - E^s - g(K^s, f(K,L)) = 0 \tag{42}$$

$$\lambda: s[K + K^s] - P(f(K,L))f(K,L) + wL + c^+E^p - c^-E^s \geqq 0 \tag{43}$$

$$\lambda[s[K + K^s] - P(f(K,L))f(K,L) + wL + c^+E^p - c^-E^s] = 0 \tag{44}$$

$$K, L, K^s, E^p, E^s, \gamma, \delta, \lambda \geqq 0 \tag{45}$$

These conditions can be used to contrast the regulated firm's method of meeting the emission standard with the unregulated firm's. Rather than go through all eight possibilities, two of the most important are highlighted: namely possibilities 2 ($K^s > 0$, $E^p > 0$) and 3 ($K^s > 0$ and $E^s > 0$). For these possibilities, the following proposition holds.

Proposition: for a firm that faces a rate-of-return constraint and that either installs scrubbers ($K^s > 0$) and purchased entitlements ($E^p > 0$) or that installs scrubbers and sells entitlements ($E^s > 0$) to satisfy its emission standard, more scrubbers are installed and fewer entitlements purchased or more entitlements sold relative to a firm without a constraint on its rate-of-return that installs scrubbers and purchases or sells entitlements.

Proof of this proposition is in the appendix. The point to be made is that a rate-of-return constraint biases the firm in the direction of greater capital intensity. By building up a greater capital base on which a return can be made, profits can be increased. Relative to the firm operating without a regulatory constraint, at the margin this firm finds scrubbers more attractive than entitlements for meeting the emission standard. Both entitlements and scrubbers are costly but scrubbers also increase the rate base. As the appendix shows, at the optimum for possibility 2 and for possibility 3

$$c^+ < -\frac{r}{g_s} \tag{46}$$

$$c^- < -\frac{r}{g_s} \tag{47}$$

The firm is not responding to the correct price and cost signals in making its selection. Recall from Equations 24 and 25 that equality between the marginal cost of entitlements or marginal revenue from entitlements and the marginal cost of scrubbers is necessary if the standard is to be met at minimum cost. The rate-of-return firm purchases scrubbers to the point where their marginal cost exceeds either the marginal cost of entitlements or the marginal revenue from entitlements. Moreover, this inefficiency is spread through the economy. If the firm is purchasing entitlements, and it falls short of the correct number of purchased entitlements, then other firms are not selling as many entitlements as they would in the absence of rate-of-return regulation. These other firms, therefore, must meet their standards with fewer scrubbers and more entitlements. Yet their scrubbers may be less expensive than the rate-of-return firm's scrubbers, and the result is too many expensive scrubbers and not enough inexpensive scrubbers utilized. The rate-of-return firm selling entitlements sells too many, because it can cover the resulting deficit in the emission standard with scrubbers, which are attractive as rate base builders. This again implies that the cost across all firms is not being miminized, since other firms are buying too many entitlements.

The proposition crucially depends upon K^s being included in the firm's rate base. Eliminating K^s from the rate base would yield a bias in the opposite direction; that is, fewer scrubbers and greater participation in the entitlement system than is optimum. Since the timing of including scrubbers in the rate base varies among states as indicated above, the bias toward greater scrubbers may also vary. In states where scrubbers are included in the rate base during construction, a stronger bias toward the use of scrubbers should be exhibited than in states where scrubbers are included after construction. This hypothesis could be shown theoretically by adding a time dimension to the model herein, and it also lends itself to empirical investigation.

OTHER COMPLICATIONS FOR TDP

The second reason for violations of Montgomery's profit maximization assumption is based in a large body of literature concerning why firms do not or cannot pursue this goal. "The argument, in brief, is that profit maximization is at best unappealing and at worst meaningless to business decisionmakers operating in an environment of dynamic uncertainty, organizational complexity, and conflicting goals" [13]. Taking these in order, uncertainty clearly plays an important role in the operations of any firm. What will future demand for its products be? What will wage rates be after the next labor contract expires? Are interest rates and capital costs falling or rising? What are rival sellers planning? The list is endless, yet management must still make decisions. Usually, the firm is characterized as recognizing these random events, and able to decipher information about the distributions of such events. Then decisions can be based on expected values of interest rates, wages, etc. However, nearly identical managers in identical situations may make very different decisions under uncertainty, and neither need be wrong. The differences may lie in attitudes toward risk. For example, given nearly identical firms, one may install pollution control equipment and the other may opt for purchasing entitlements to meet emission standards. Installation may take several years over which time interest rates can change and the cost of the equipment is uncertain. The firm choosing to install either has a lower subjective estimation about future interest rates, or has a similar estimate but is less risk averse towards the capital expenditure. Are both firms then obeying rules of the type dictated above where marginal costs are equated? Firms make different subjective judgments about the cost of capital, r, and are consequently making decisions based on different signals. Yet only one true cost of capital will prevail. Under these conditions, the attainment of a least cost solution to emissions control is uncertain, and perhaps even undefinable.

Organizational complexity refers to the idea that modern firms can be so large and have such complicated hierarchical structures that management may simply be unable to maximize profits. Responsibility for a firm's operations must be handled by specialists in finance, production, sales, accounting, research and development, and so on. Firms that are heavy polluters may have separate departments specializing in meeting emissions standards. Also, a communications network must encompass all these various departments and the larger the network, the greater the chance that information flows will be distorted. Given this complexity, is it reasonable to expect profit-maximizing decisions? Even in the simple model presented earlier, there needs to be close cooperation between the production decisions where K and L are joined to obtain output, the finance decision that must consider the cost of capital including scrubbers K^s and the pollution control decision that must decide between installing scrubbers and purchasing or selling entitlements. Teamwork for all these decisions is required if the appropriate marginal costs are to be equated as per Equations 24 and 25. However, incorrect or misinterpreted instructions flowing between departments or managers that have their own personal goals in mind may contribute to violations of Equations 24 and 25 and non-cost-minimizing solutions to meeting standards.

Finally, there is the possibility that firms may have conflicting goals or at least goals other than profit maximization. Simon [14] has labeled this phenomena satisfying. For example, a firm may be interested in profits, but after attaining some particular profit level, other goals come into play. Two other goals are considered here. First, sales maximization may be an alternative goal if managers associate their salaries and prestige with sales volume. In this case, the objective function for the firm is simply p(y)y, although the same emissions constraint given by Equation 4 is imposed. The problem is that there is no mechnanism now to choose between scrubbers and entitlements. More scrubbers or more entitlements will both allow more emissions and, therefore, a greater sales volume; the incentive to make a cost minimizing choice between scrubbers and entitlements is lacking. While profits are affected by the choice, sales are not.

A second alternative goal is staff maximization. Here, management's prestige may be linked with the size of the work force. Consider a management that desires a large staff but is also aware of the importance of profit. Then a multivariate utility function may be appropriate; that is, $U(\pi,L)$ may be a management utility function illustrating that managers view both profit and labor as providing personal satisfaction. A tradeoff then exists between profits and staff, and the firm will be biased towards a larger work force than the profit-maximizing firm. How this effects attaining emission standards depends on whether operating scrubbers is more or less labor-intensive than dealing in TDP markets. Whichever is more intensive will be the favored solution, and again Equations 24 and 25 will not be realized.

SUMMARY AND CONCLUSION

A model of a profit-maximizing firm was presented where the firm confronts an emission standard and has the options to either install scrubbers or enter into TDP markets. The results of the model characterize how the firm chooses among the options to meet the standard at minimum cost. However, the driving force behind the minimum cost solution is that the firm is a profit maximizer. There are situations where this may not be an appropriate assumption. For instance, in regulated industries where constraints on profits are invoked by government commissions, the minimum cost solution is not attained. A bias toward overutilization of scrubbers is possible. Even in nonregulated industries, however, there are plausible reasons why firms do not maximize profit, including uncertainty, organization complexity and alternative goals.

All of this is not to say that TDP markets should be abandoned. Instead, this chapter contains a number of pitfalls that regulators should be aware of when implementing TDP markets. How important these pitfalls are is an empirical question that cannot be answered until the markets are allowed to operate for some time. Until then, control agencies should closely observe those firms that may be candidates for these pitfalls, particularly those in regulated industries. And if their behavior deviates from that outlined above, adjustments in the TDP system are in order.

APPENDIX

In proving the proposition, g_s is assumed to be an increasing function in K^s, or $g_{ss} > 0$. This implies that the marginal cost of scrubbers, $-r/g_s$ is increasing, or that the cost of removing the nth unit of emission is greater than the cost of removing the $n - 1$ unit. This is a standard assumption and is consistent with second order sufficient conditions for a profit maximum.

Returning to the proposition, Equations 33 through 36 for $K^s > 0$, $E^p > 0$, and $\delta = 0$ imply

$$c^+ + \frac{r}{g_s} = \lambda \left(c^+ + \frac{s}{g_s} \right) \tag{A-1}$$

The problem is to show that Equation A-1 is negative to confirm Equation 46. First, λ is assumed to be positive and the rate-of-return constraint binding. From Equation 35 as an equality,

$$c^+(1 - \lambda) = \gamma \tag{A-2}$$

Since $\gamma \geqq 0$ for a binding emission standard, it must be that

$$0 < \lambda \leqq 1 \tag{A-3}$$

Now suppose Equation A-1 is positive, then Equation A-3 implies

$$c^+ + \frac{r}{g_s} \leqq c^+ + \frac{s}{g_s}$$

or

$$\frac{r}{g_s} \leqq \frac{s}{g_s}$$

which is a contradiction since $g_s < 0$ and $s > r$. Consequently, Equation A-1 is negative, $c^+ + r/g_s < 0$ and Equation 46 follows. Since the non-rate-of-return firm satisfies Equation 24 and since $g_{ss} > 0$, it follows that g_s is greater for the rate-of-return firm and more scrubbers are purchased. Using Equations 33, 34, 37 and 38, an identical argument can be used to show Equation 47 for $K_s > 0$ and $E^s > 0$.

REFERENCES

1. Crocker, T.D. "The Structuring of Atmospheric Pollution Control Systems," in *The Economics of Air Pollution,* Harold Wolozin, Ed. (New York: W.W. Norton, 1966).
2. Dales, J.H. *Pollution, Property, and Prices* (Toronto: University of Toronto Press, 1968).
3. Tietenberg, T.H. "Transferable Discharge Permits and the Control of Stationary Source Air Pollution: A Survey and Synthesis," *Land Econ.* 56(4):391–419 (1980).
4. Atkinson, S.E., and T.H. Tietenberg. "The Empirical Properties of Two Classes of Designs for Transferable Discharge Permit Markets," *J. Environ. Econ. Manage.* (in press).
5. "Emission Reduction Banking Manual," PM-220 U.S. Environmental Protection Agency (1980).
6. Montgomery, D.W. "Markets in Licenses and Efficient Pollution Control Programs," *J. Econ. Theory* 5:395–418 (1972).
7. Averch, H., and L.L. Johnson. "Behavior of the Firm Under Regulatory Constraint," *Am. Econ. Rev.* 52:1052–1069 (1962).
8. Takayama, A. "Behavior of the Firm Under Regulatory Constraint," *Am. Econ. Rev.* 59:255–260 (1969)
9. Spann, R.M. "Rate of Return Regulation and Efficiency in Production: An Empirical Test of the Averch-Johnson Thesis," *Bell J. Econ.* 5(1):38–52 (1974).
10. Courville, L. "Regulation and Efficiency in the Electric Utility Industry," *Bell J. Econ.* 5(1):53–74 (1974).
11. Boyer, W.J. "An Empirical Examination of the Averch-Johnson Effect," *Econ. Inquiry* 14(1):25–35 (1976).
12. Berg, S. "Strategic Responses to Environmental Regulation: Long Range Transport of Air Pollution," prepared for the Office of Technology Assessment, U.S. Congress (1981).
13. Sherer, F.M. *Industrial Market Structure and Economic Performance,* 2nd ed. (Chicago: Rand-McNally, 1980).
14. Simon, H. "Theories of Decision-Making in Economics and Behavioral Science," *Am. Econ. Rev.* 49:253–283 (1959).

INDEX